A Study
of
Bernhard Riemann's 1859 Paper

by Terrence P. Murphy

ISBN No: **978-0-9961671-3-0**
Library of Congress Control Number: **2020916158**

Publisher:
Paramount Ridge Press
Bellevue, WA 98008
terry@riemann1859.com

The formatting/layout of this book was produced using MiKTEX (with the LuaTEX compiler), which is a software front-end for the TEX / LATEX typesetting and document preparation system.

The fonts used in this book are:

Serif font (primary document font): **TeX Gyre Termes**
Math font: **TeX Gyre Termes Math**
Sans Serif font: **Latin Modern Sans**
All fonts: © Bogusław Jackowski and Janusz M. Nowacki
Licensed under version 1.0 of the GUST Font License: http://www.gust.org.pl/projects/e-foundry/licenses

Front cover by Terrence P. Murphy
Book design by Terrence P. Murphy
Printed in the United States of America.

Contents

Preface

The primary purpose of this book is to deeply study Bernhard Riemann's seminal 1859 paper: *On the Number of Primes Less Than a Given Magnitude* [31] ("Riemann's Paper"). Certainly, the most influential paper in the history of analytic number theory, Riemann's Paper contained concepts so advanced that it took more than 30 years for the greatest mathematicians of his day to fully grasp the ideas advanced in his paper.

Today, Riemann's Paper is famous for the "Riemann Hypothesis" – a brief comment that Riemann made regarding the location of the zeros of the zeta function. This book, however, is focused on the rest of Riemann's Paper. And for good reason. To begin to understand the Riemann Hypothesis, you must first understand all of Riemann's Paper. We hope this book will help you do just that.

It is important to separate the *ideas* in Riemann's Paper from the actual text. Reading Riemann's Paper is, to say the least, difficult, as well-described below[1]:

> *[Riemann's Paper]...is written in an extremely terse and difficult style, with huge intuitive leaps and many proofs omitted. This led to (in retrospect quite unfair) criticism by Landau and Hardy in the early 1900's, who commented that Riemann had only made conjectures and had proved almost nothing. The situation was greatly clarified in 1932 when Siegel [cite] published his paper, representing about two years of scholarly work studying Riemann's left over mathematical notes at the University of Göttingen, the so-called* Riemann's Nachlass. *From this study it became clear that Riemann had done an immense amount of work [related to his paper] that never appeared in his paper. One conclusion is that many formulae that lacked sufficient proof in [his paper] were in fact proved in these notes. A second is that the notes contained further discoveries of Riemann that were never even written up in [his paper]. One such is what is now called the Riemann-Siegel formula...*

Our goal in this book is to provide rigorous proofs for all of the proofs and (provable) assertions in Riemann's Paper. Of course, that necessarily excludes the Riemann Hypothesis.

While Riemann's Paper is our focus, our study would be incomplete without also noting some of the advances made as a result of his paper. Most notably, we provide two proofs of the *Prime Number Theorem*.

Who Is This Book For?

If you are reading this, chances are you have developed a keen interest in the Riemann Hypothesis. Maybe you read John Derbyshire's excellent book *Prime Obsession* [12]. Or perhaps you read that the Riemann Hypothesis is one of the seven *Millennium Prize Problems*, as selected by the Clay Mathematics Institute, with a $1 million prize for its proof.

This book is intended to bring an understanding of Riemann's paper to a wider audience. Our goal is to bridge the gap between Derbyshire's excellent but less technical book, and Harold Edward's excellent but highly technical book, *Riemann's Zeta Function* [13].

To advance your knowledge beyond Derbyshire's book, you must have a good understanding of complex analysis (call it knowledge at the "hobbyist" level). By comparison, you need to be a complex analysis "guru" to understand Edward's book. The key point: if you are not at the hobbyist level, this book is *probably not for you*.

Several people have told me they want to get to the hobbyist level so they can better understand this book. However, when they look at Ahlfors and other well-known (400+ page) textbooks on complex analysis, it feels like a mountain just a little too high to climb. In that case, you might consider my companion book *Complex Analysis: a Self-Study Guide* [25].

What Is In This Book?

Chapters 1 through 3: The Setup

Chapter 1: Prime Numbers. Prime numbers are at the heart of Riemann's Paper. We start at the beginning – Euclid's proof that there are infinitely many prime numbers. After discussing the seeming randomness of primes numbers, we note the intense interest in the early 1800's in finding a formula that provides a "really good" estimate of the number of primes less than any (large enough) number – that is, the *Prime Number Theorem*.

Chapter 2: The Harmonic Series (to the Zeta Function). Prime numbers are at the center Riemann's Paper, but the zeta function is the key that unlocks their secrets. Here we follow the development of the zeta function during the period before Riemann's Paper, particularly the important work of Leonhard Euler.

Chapter 3: The Factorial (to the Gamma Function). We next follow the development of the gamma function, and again see that Leonhard Euler is a central player. Later in this book, you will see that the full power of the zeta function cannot be unleashed without the critical assistance of the gamma function.

Chapters 4 through 11: Riemann's Paper and Proofs

Chapter 4: Riemann's 1859 Paper. We present here the complete text of Riemann's Paper. We divide Riemann's Paper into sections, annotate each section, and give a reference forward to the chapter(s) in this book where we study that section of his paper.

Chapter 5: Zeta – Analytic Continuance. We prove the analytic continuation of $\zeta(s)$ (the zeta function) to the entire complex plane, with one simple pole at $s = 1$. We also include an elementary proof of the analytic continuation of $\zeta(s)$ to the right half-plane ($Re(s) > 0$).

Spot Any Errors?

If you find any typographical or other errors in this book, the author would very much appreciate hearing from you. Drop a note to terry@riemann1859.com. Other feedback (e.g., some proof/concept is not very well explained) is also appreciated. Our plan is to list *errata* at www.riemann1859.com.

Notation

$\displaystyle\sum_{\rho} \quad \prod_{\rho}$ The sum / product taken over all roots ρ of ξ (that is, all roots of ζ in the critical strip), with their multiplicities and in order of increasing $|Im(\rho)|$. In the text, we will sometimes require that ρ and $1 - \rho$ be paired (included together in a single term of the sum / product).

$\displaystyle\sum_{p} \quad \prod_{p}$ The sum / product taken over the ordered set of all prime numbers.
Note the difference: The letter "p" is used here for primes. The Greek letter "ρ" is used (above) for the roots ξ. Although the letters look very similar, the context should always make clears which one is intended.

$\displaystyle\sum_{[[n=x]]} f(n)$ The sum is $f(x)$ if $n = x$, and 0 otherwise.

$[x]$	The integral part of the real number x.
$\{x\}$	The fractional part of the real number x.
$\log(x)$	The natural logarithm $\log_e(x) = \ln(x)$.
$Re(z)$	The real part of the complex number z.
$Im(z)$	The imaginary part of the complex number z.
$f(z) = \mathcal{O}(g(z))$	There exists a constant $C > 0$ such that $|f(z)| \le C|g(z)|$ for sufficiently large $|z|$.
$\zeta(s)$	The *Riemann Zeta Function*, as defined in Chapter 2 and Chapters 5 through 7.
$\xi(s)$	The *Completed Zeta Function*, as defined in Chapter 8.
$\mathbb{N}$	The set of positive integers: $1, 2, 3, ...$
$\mathbb{N}_{\ge 0}$	The set of non-negative integers: $0, 1, 2, 3, ...$
$\mathbb{Z}$	The set of all integers: $..., -3, -2, -1, 0, 1, 2, 3, ...$
$\mathbb{R}$	The set of real numbers.
$\mathbb{C}$	The set of complex numbers.
critical strip	The area in the complex plane $\{s \in \mathbb{C} : 0 \le Re(s) \le 1\}$.
LHS or RHS	The left-hand side (or right-hand side) of a given equation.
Let (or fix) $a > 1$	For any variable (e.g., a) defined with an inequality, it is assumed $a \in \mathbb{R}$.

Chapter 1

Prime Numbers

1.1 Euclid's *Elements*

Prime numbers are at the center of this book. They have fascinated mathematicians for more than two thousand years and have still not revealed all of their secrets.

Even today, much of what we know about prime numbers was known at the time of the great Greek mathematician, Euclid (323BC to 283BC). A full treatment of prime numbers was included in his broad-ranging book on mathematics, *Elements*, no doubt the most influential book in the history of mathematics. It served as a main textbook for teaching mathematics (especially geometry) for over two millennia, from the time of its publication (about 300BC) until the 18th century.

Three of the key items contained in *Elements* are the definition of prime numbers, the fundamental theorem of arithmetic and the proof of the infinitude of prime numbers. All are discussed below. But there were several other references to primes in *Elements*. For example, for prime p, Euclid showed that if $2^p - 1$ is a prime (now called a Mersenne prime), then $2^{p-1}(2^p - 1)$ is a perfect number (a positive integer that is equal to the sum of its proper positive divisors, excluding the number itself).

1.2 Prime Number Defined

A prime number is a natural number greater than 1 which cannot be expressed as the product of two smaller natural numbers other than 1 and itself. A natural number greater than 1 that is not a prime number is a composite number. For example, 5 is prime because 1 and 5 are its only positive integer divisors, whereas 6 is composite because it has the divisors 2 and 3 in addition to 1 and 6.

Following are all prime numbers less than 200: 2, 3, 5, 7, 11, 13, 17, 19, 23, 29, 31, 37, 41, 43, 47, 53, 59, 61, 67, 71, 73, 79, 83, 89, 97, 101, 103, 107, 109, 113, 127, 131, 137, 139, 149, 151, 157, 163, 167, 173, 179, 181, 191, 193, 197, 199.

1.3 The Fundamental Theorem of Arithmetic

The fundamental theorem of arithmetic states that: (1) every natural number larger than 1 is either a prime number or is the product of two or more primes numbers, and (2) this product is unique up to rearrangement. For example:

$$1200 = 2^4 \times 3^1 \times 5^2$$

Under the theorem, 1200 can be represented as a unique product of primes. When we say that product is unique, we mean there will always be exactly four 2's, one 3, two 5's, and no other primes in the product.

Note that if p is a prime number and p divides a product ab of integers, then p divides either a or b (or both).

1.4 There Are Infinitely Many Prime Numbers

Following is Euclid's proof of the infinitude of primes:

Theorem 1.1. *Prime numbers are more than any assigned multitude of prime numbers.*

Proof. Suppose that there are n primes, $p_1, p_2, ..., p_n$. Let K be the product of $p_1, p_2, ..., p_n$. Now consider the number $K + 1$.

If $K + 1$ is prime, then there are at least $n + 1$ primes.

If $K + 1$ is not prime, then some prime p_j divides it. But p_j cannot be any of the primes $p_1, p_2, ..., p_n$, because they all divide K and do not divide $K + 1$. Therefore, there are at least $n + 1$ primes. $\square$

1.5 The Distribution of Primes...In So Many Words

It is hard to make sense of the distribution of primes. They pop up seemingly at random, with no rhyme or reason. We devote this section to the words of several famous mathematicians describing the confoundedness of prime numbers:

Andrew Granville (1997):

> *Prime numbers are the most basic objects in mathematics. They also are among the most mysterious, for after centuries of study, the structure of the set of prime numbers is still not well understood. Describing the distribution of primes is at the heart of much mathematics...*

In a letter from John Littlewood to G.H. Hardy regarding a third mathematician (circa 1900):

> *It is not surprising that he would have been [misled], unsuspicious as he presumably is of the diabolical malice inherent in the primes.*

Leonhard Euler (1751):

> *Since primes are the basic building blocks of the number universe from which all the other natural numbers are composed, each in its own unique combination, the perceived lack of order among them looks like a perplexing discrepancy in the otherwise so rigorously organized structure of the mathematical world. . . . How can so much of the formal and systematic edifice of mathematics, the science of pattern and rule and order per se, rest on such a patternless, unruly, and disorderly foundation? Or how can numbers regulate so many aspects of our physical world and let us predict some of them when they themselves are so unpredictable and appear to be governed by nothing but chance?*

Don Zagier from "The first 50 million prime numbers", (1977):

There are two facts about the distribution of prime numbers which I hope to convince you so overwhelmingly that they will be permanently engraved in your hearts. The first is that despite their simple definition and role as the building blocks of the natural numbers, the prime numbers... grow like weeds among the natural numbers, seeming to obey no other law than that of chance, and nobody can predict where the next one will sprout. The second fact is even more astonishing, for it states just the opposite: that the prime numbers exhibit stunning regularity, that there are laws governing their behavior, and that they obey these laws with almost military precision.

P.J. Davis and R. Hersh, The Mathematical Experience, Chapter 5:

Some order begins to emerge from this chaos when the primes are considered not in their individuality but in the aggregate; one considers the social statistics of the primes and not the eccentricities of the individuals.

Marcus du Sautoy, The Music of the Primes (2003):

Armed with his prime number tables, Gauss began his quest. As he looked at the proportion of numbers that were prime, he found that when he counted higher and higher a pattern started to emerge. Despite the randomness of these numbers, a stunning regularity seemed to be looming out of the mist.

The revelation that the graph appears to climb so smoothly, even though the primes themselves are so unpredictable, is one of the most miraculous in mathematics and represents one of the high points in the story of the primes. On the back page of his book of logarithms, Gauss recorded the discovery of his formula for the number of primes up to N in terms of the logarithm function. Yet despite the importance of the discovery, Gauss told no one what he had found. The most the world heard of his revelation were the cryptic words, "You have no idea how much poetry there is in a table of logarithms".

1.6   Tracking Down the Primes

Since ancient Greek times, mathematicians have developed algorithms for finding the prime numbers up to any given limit. The unfortunate truth is that all such algorithms are computationally expensive. Even today, you can't do all that much better than the Sieve of Eratosthenes, named after Eratosthenes of Cyrene (276BC to 194BC):

- **Goal:** find all the prime numbers less than or equal to a given integer n.
- **Step 1:** Create a list of consecutive integers from 2 through n: $(2, 3, 4, ..., n)$.
- **Step 2:** Initially, let p equal 2, the smallest prime number.
- **Step 3:** Enumerate the multiples of p by counting in increments of p from $2p$ to n, and mark them in the list (these will be $2p, 3p, 4p, ...$; the p itself should not be marked).
- **Step 4:** Find the first number greater than p in the list that is not marked. If there was no such number, go to step 5. Otherwise, let p now equal this new number (which is the next prime), and repeat from step 3.

- **Step 5:** The remaining unmarked numbers in the list are all the primes below n.

Some later improvements include:

- The step 1 list of integers can exclude all even integers greater than 2.
- In step 3, you can skip all multiples of p less than p^2.
- In step 4, go to step 5 if $p^2 > n$.

Note also that all prime numbers except 2 and 3 are of the form $(6n \pm 1)$ for some integer n. To see that, divide an integer by 6. If the remainder is 0, 2 or 4, the number is even. If the remainder is 3, the number is divisible by 3.

1.7 The Still Unknown

Even today, many questions regarding prime numbers remain open. We list just a *very small* sampling here:

Goldbach's Conjecture: every even integer greater than 2 can be expressed as the sum of two primes.

The Twin Prime Conjecture: there are infinitely many pairs of primes whose difference is 2.

Legendre's conjecture: there is a prime number between n^2 and $(n + 1)^2$ for every positive integer n.

Polignac's conjecture: for every positive integer n, there are infinitely many pairs of consecutive primes that differ by $2n$.

One of Landau's problems: there are infinitely many primes of the form $n^2 + 1$.

H. Brocard's conjecture: there are always at least four primes between the squares of consecutive primes greater than 2.

1.8 The Prime Number Theorem

Having focused above on the seeming unpredictability and randomness of primes, we now turn away from individual prime numbers (rogues all) and turn to the statistical behavior of primes in the large.

We consider the following question. Given some positive integer N, how many prime numbers are there between 2 and N? There is no function that takes N as its input and gives you the number of primes between 2 and N as its output. The only way to answer the question for any N is to do an actual count. The above discussion regarding the distribution of primes should convince you of that fact.

Having admitted defeat on finding an exact prime counting function, mathematicians did not throw in the towel. They set as their goal the next best thing: a "really, really good" estimating function. We describe the concept of a prime count estimating function below.

1.8.1 Prime Count Estimating Function

We know that an exact prime counting function does not exist. We don't know (at this point in this book) whether a "really, really good" prime count estimating function exists. That does not stop us from thinking of two magic functions that do just that. So, bear with me as we define:

$A(N)$: the function A (A for "Actual") takes N as its input and outputs the exact number of primes between 2 and N. (Of course, this magic function does not really exist).

$E(N)$: the function E (E for "Estimate") takes N as its input and, using some formula, outputs a "really, really good" estimate of the number of primes between 2 and N.

We must first define "really, really good" in mathematical terms.

1.8.2 Asymptotically Equivalent

In mathematics, we say the A and E functions are "asymptotically equivalent" if their ratio gets arbitrarily close to one-to-one if N is large enough:

$$\lim_{N \to \infty} \frac{A(N)}{E(N)} = 1.$$

Obviously, if $A(N)$ divided by $E(N)$ exactly equals 1, then $A(N) = E(N)$ and we can say that $E(N)$ is exactly equal to the prime counting function. But that is not what the above formula means and that is not what asymptotically equivalent means.

Think of it as a challenge and response test. You challenge me by giving me a very small but non-zero positive number. For example, you give me the challenge number of one-trillionth $(0.0000000000001) = 1/10^{12}$. Then, I must respond with a number M such that:

$$\left(1 - \frac{1}{10^{12}}\right) \le \frac{A(N)}{E(N)} \le \left(1 + \frac{1}{10^{12}}\right) \text{ for all } N > M.$$

In other words, the ratio of $A(N)/E(N)$ is within one-trillionth of 1 for all N greater than some M.

So, two functions are asymptotically equivalent if, for any small (but non-zero) positive number you give me, I can find a number M such that (for all N greater than M) the ratio of the two functions is within your small number of 1. That is what we mean by a "really, really good" estimate.

1.8.3 Ideas Emerge for an Estimating Function

By the 1790's, prime number tables were being created that included larger and larger prime numbers. As the numbers got larger, patterns were emerging from the chaos. Some of the first attempts were not well considered. But it did not take long before a consensus developed around some variation of the following:

$$\pi(N) \sim \frac{N}{log(N)}.$$

Note that π is the proper mathematics lingo for the prime counting function (i.e., the same theoretical function as our $A(N)$ in the discussion above). Note also that the $\sim$ means asymptotically equivalent. So, if the mathematicians of that day were correct, then our estimating function $E(N)$ consists of a simple formula: it takes N as its input and outputs N divided by $log(N)$, where $log(N)$ is the natural (base e) logarithm function.

1.8.4 Statement of The Prime Number Theorem

With all the unpredictability and randomness of prime numbers, could the prime count estimating function be that simple? Putting the above together, the *Prime Number Theorem* would state:

$$\lim_{N \to \infty} \frac{A(N)}{E(N)} = \lim_{N \to \infty} \frac{\pi(N)}{\left(\frac{N}{log(N)}\right)} = 1.$$

Stating the theorem is a long way from proving it. The first roadblock, obviously, is that there is no such function as $\pi(N)$. It is a conceptual/theoretical function, but there is simply no function that gives us the exact number of prime numbers between 2 and N. (Put another way, we would not need an estimating function if we had the real thing.)

So, how do you go about proving the above two functions (one theoretical and one real) are asymptotically equivalent? It would take about 100 years from first conception to actual proof. And that proof would rely in large part on the ideas developed in Riemann's Paper.

Chapter 2

The Harmonic Series (to the Zeta Function)

We begin this chapter with a review of the Harmonic Series, a mildly interesting infinite series, well-known even in the 14th century. Next, we study the Basel Series, an interesting variation of the Harmonic Series, perhaps most famous for making Leonhard Euler (the greatest mathematician of the 18th century) famous.

From there we move to the real-valued zeta function, which generalizes the Harmonic Series and the Basel Problem into a class of similar functions. As so defined, the zeta function might be considered interesting, but certainly not important. But then Leonhard Euler (yet again) pulls a rabbit out of his hat. He found another representation for the zeta function, known as the *Euler Product Formula*. That formula placed the zeta function at the absolute center of the quest for a greater understanding of prime numbers.

2.1 The Harmonic Series in Music

Pythagoras (570BC to 495BC) was a great mathematician who also had a deep interest in music and harmony. The story goes that he one day passed a metals shop and noticed workmen pounding out pieces of metal on anvils. He took note of the differences in sound and pitch between larger and smaller hammers. Later, experimenting at home with stringed musical instruments, he noted the relationship between string length and pitch. He is credited with developing the first scientific approach to musical concepts of pitch and harmony. In fact, prior to the 1500's, his system of musical tuning ("Pythagorean tuning") was in common use.

Although the topic is well beyond the scope of this book, in brief, a harmonic series in music is the sequence of sounds where the base frequency of each sound is an integer multiple of the lowest base frequency. In addition to vibrating over its entire length, a string simultaneously vibrates over fractional divisions of its length (1/2, 1/3, 1/4, 1/5, 1/6, etc.) producing a series of harmonics (overtones) whose frequencies are inversely proportional.

2.2 The Harmonic Series in Mathematics

In mathematics, the harmonic series is the following infinite series:

$$\sum_{n=1}^{\infty} \frac{1}{n} = \frac{1}{1} + \frac{1}{2} + \frac{1}{3} + \frac{1}{4} + \ldots$$

After the above discussion of the harmonic series in music, it is immediately clear why the mathematical series is known as the harmonic series.

What is the sum of this infinite series of fractions? Does it "converge" (the sum is less than some finite number) or "diverge" (the sum is greater than any finite number)?

For the harmonic series, you can see that the nth term of the series is shrinking rapidly towards zero. In fact, the sum of the first 10^{43} terms is less than 100. In case you are wondering, the term that gets the sum over 100 is:

$$\frac{1}{15,092,688,622,113,788,323,693,563,264,538,101,449,859,497}$$

So, does the harmonic series converge or diverge? Given the above facts, most would guess that the series converges. They would be wrong.

2.3 Two Proofs that the Harmonic Series Diverges

Proof #1 – The Comparison Test. This first proof of divergence was in 1360 by Nicole Oresme (1325 to 1382). We compare the harmonic series (the first below) to another series as follows:

$$1 + \frac{1}{2} + \left(\frac{1}{3} + \frac{1}{4}\right) + \left(\frac{1}{5} + \frac{1}{6} + \frac{1}{7} + \frac{1}{8}\right) + \left(\frac{1}{9} + ...\right)$$
$$1 + \frac{1}{2} + \left(\frac{1}{4} + \frac{1}{4}\right) + \left(\frac{1}{8} + \frac{1}{8} + \frac{1}{8} + \frac{1}{8}\right) + \left(\frac{1}{16} + ...\right)$$

The number of terms in the parenthetical doubles each time we move right to the next parenthetical. The **final** term in each parenthetical sum in the harmonic series is equal to **every** term in the matching parenthetical in the second series.

But that makes the sum within each parenthetical in the harmonic series greater than its matching parenthetical sum in the second series. Which makes the harmonic series greater than the second series.

But the second series is equivalent to

$$1 + \frac{1}{2} + \frac{1}{2} + \frac{1}{2} + \frac{1}{2} + ... \to \infty.$$

So, by the comparison test, the harmonic series also diverges. □

Proof #2 – The Integral Test. This is the more modern proof (integrals were not even known at the time of Nicole Oresme). For any $N \in \mathbb{N}$, we have

$$\sum_{n=1}^{N} \frac{1}{n} = \sum_{n=1}^{N} \int_{n}^{n+1} \frac{1}{n} \, dx \geq \sum_{n=1}^{N} \int_{n}^{n+1} \frac{1}{x} \, dx = \int_{1}^{N+1} \frac{1}{x} \, dx = \log(N+1) - \log(1) = \log(N+1).$$

Therefore

$$\sum_{n=1}^{\infty} \frac{1}{n} = \lim_{N \to \infty} \sum_{n=1}^{N} \frac{1}{n} \geq \lim_{N \to \infty} \log(N+1) \to \infty. \qquad □$$

2.4 The Basel Problem

It is not much of a leap to make a variation of the harmonic series with one simple change:

$$\sum_{n=1}^{\infty} \frac{1}{n^2} = \frac{1}{1^2} + \frac{1}{2^2} + \frac{1}{3^2} + \frac{1}{4^2} + \dots$$

We start with the harmonic series but then we square each term in the series. It is well-known (and easily proved) that this series converges. But exactly what number does this series sum to?

Pietro Mengoli (1626 to 1686) was an Italian mathematician from Bologna. In 1644, Mengoli posed the above question, the famous *Basel problem*. It is called the *Basel problem* because much of the work in attempting to solve the problem was done at Basel University by members of the Bernoulli family (who were ultimately unsuccessful).

As it turned out, the problem was solved by Leonhard Euler in 1735 (made more rigorous by him in 1741). Euler's hometown was also Basel. Since the problem had withstood the attacks of the leading mathematicians of the day for 90+ years, Euler's solution brought him immediate fame at the age of 28.

We will describe Euler's solution, without presenting a formal proof. We start with the Taylor Series expansion for $\sin(x)$:

$$\sin(x) = x - \frac{x^3}{3!} + \frac{x^5}{5!} - \frac{x^7}{7!} + \frac{x^9}{9!} \dots$$

Divide by x to obtain:

$$\frac{\sin(x)}{x} = 1 - \frac{x^2}{3!} + \frac{x^4}{5!} - \frac{x^6}{7!} + \frac{x^8}{9!} \dots \tag{2.1}$$

Euler's was, of course, well aware of the *Fundamental Theorem of Algebra*. Every complex polynomial $p(z)$ of degree $n \geq 1$ has a factorization $p(z) = c(z - z_1)^{m_1} \dots (z - z_k)^{m_k}$ where the z_j's are distinct and $m_j \geq 1$. This factorization is unique, up to a permutation of the factors. For example:

$$p(z) = z^3 - 5z^2 - 4z + 20 = (z - 5)(z + 2)(z - 2).$$

As you see, $p(z)$ can be factored by its zeros: $5, -2$, and 2.

Euler's "leap" was to apply the *Fundamental Theorem of Algebra* to a function having an infinite numbers of zeros. This needed the *Weierstrass factorization theorem* (with all its subtleties and complexities), which was not proved for another 100+ years. Fortunately, the sine function was a perfect candidate for the *Weierstrass factorization theorem*, so Euler's "leap of faith" was well-rewarded.

The zeros of the sin function are found at $\pm k\pi$ for $k \in \mathbb{N}_{\geq 0}$. Euler therefore wrote:

$$\frac{\sin(x)}{x} = \left(1 - \frac{x}{\pi}\right)\left(1 + \frac{x}{\pi}\right)\left(1 - \frac{x}{2\pi}\right)\left(1 + \frac{x}{2\pi}\right)\left(1 - \frac{x}{3\pi}\right)\left(1 + \frac{x}{3\pi}\right) \dots$$

$$= \left(1 - \frac{x^2}{\pi^2}\right)\left(1 - \frac{x^2}{4\pi^2}\right)\left(1 - \frac{x^2}{9\pi^2}\right) \dots$$

To convert the infinite product back to an infinite series, Euler now multiplied and collected the x^2 terms:

$$-\left(\frac{1}{\pi^2} + \frac{1}{4\pi^2} + \frac{1}{9\pi^2} \cdots\right) = \frac{-1}{\pi^2} \sum_{n=1}^{\infty} \frac{1}{n^2}.$$

Euler noticed that the coefficient of the x^2 term in equation (2.1), was $-\frac{1}{6}$, and concluded the two derivations of the coefficient must be equal:

$$-\frac{1}{6} = \frac{-1}{\pi^2} \sum_{n=1}^{\infty} \frac{1}{n^2}.$$

Multiplying both sides by $-\pi^2$ gives Euler's solution to the Basel problem:

$$\sum_{n=1}^{\infty} \frac{1}{n^2} = \frac{\pi^2}{6}.$$

There was some criticism that Euler's "proof" was a bit loose. Euler later revisited the problem and provided a tighter proof. In his later proof, the starting point was the following observation:

$$\sum_{n=1}^{\infty} \frac{1}{n^2} = \sum_{n=1}^{\infty} \frac{1}{(2n)^2} + \sum_{n=0}^{\infty} \frac{1}{(2n+1)^2}.$$

That is, the Basel Problem can be divided into two sums, a sum of even divisors and a sum of odd divisors. Then Euler notes (for the even divisors):

$$\sum_{n=1}^{\infty} \frac{1}{(2n)^2} = \sum_{n=1}^{\infty} \frac{1}{4n^2} = \frac{1}{4} \sum_{n=1}^{\infty} \frac{1}{n^2}.$$

But that means (for the odd divisors):

$$\sum_{n=0}^{\infty} \frac{1}{(2n+1)^2} = \frac{3}{4} \sum_{n=1}^{\infty} \frac{1}{n^2},$$

so the problem is reduced to proving the following equation for the odd divisors:

$$\boxed{\sum_{n=0}^{\infty} \frac{1}{(2n+1)^2}} = \frac{3}{4} \sum_{n=1}^{\infty} \frac{1}{n^2} = \frac{3}{4} \frac{\pi^2}{6} = \boxed{\frac{\pi^2}{8}}.$$

We now depart from Euler and show one (of many) modern proofs[2].

Lemma 2.1. *We prove the following:*

$$\sum_{n=0}^{\infty} \frac{1}{(2n+1)^2} = \frac{\pi^2}{8}.$$

Proof. We use the power series for the inverse sine function (valid for $|x| \leq 1$):

$$\sin^{-1}(x) = \arcsin(x) = \sum_{n=0}^{\infty} \frac{(2n-1)!!}{(2n)!!} \frac{x^{2n+1}}{(2n+1)} = x + \left(\frac{1}{2}\right)\frac{x^3}{3} + \left(\frac{1 \cdot 3}{2 \cdot 4}\right)\frac{x^5}{5} + \left(\frac{1 \cdot 3 \cdot 5}{2 \cdot 4 \cdot 6}\right)\frac{x^7}{7} + \cdots$$

where the double-factorial means the odd or even factorial (i.e., only odd or even numbers), with the convention that $-1!! = 0!! = 1$. Let $x = \sin(t)$ for $|t| \leq \pi/2$ to obtain:

$$t = \sum_{n=0}^{\infty} \frac{(2n-1)!!}{(2n)!!} \frac{\sin^{2n+1}(t)}{(2n+1)}.$$

Now use the following formula for the integral of sine to a power:

$$\int_0^{\pi/2} \sin^{2n+1}x\, dx = \frac{(2n)!!}{(2n-1)!!(2n+1)}$$

to obtain

$$\int_0^{\pi/2} t\, dt = \int_0^{\pi/2} \sum_{n=0}^{\infty} \frac{(2n-1)!!}{(2n)!!} \frac{\sin^{2n+1}(t)}{(2n+1)}\, dt = \sum_{n=0}^{\infty} \int_0^{\pi/2} \frac{(2n-1)!!}{(2n)!!} \frac{\sin^{2n+1}(t)}{(2n+1)}\, dt$$

$$= \sum_{n=0}^{\infty} \frac{(2n-1)!!}{(2n)!!(2n+1)} \int_0^{\pi/2} \sin^{2n+1}(t)\, dt = \sum_{n=0}^{\infty} \left[\frac{(2n-1)!!}{(2n)!!(2n+1)}\right]\left[\frac{(2n)!!}{(2n-1)!!(2n+1)}\right]$$

$$= \boxed{\sum_{n=0}^{\infty} \frac{1}{(2n+1)^2}} = \int_0^{\pi/2} t\, dt = \frac{(\pi/2)^2}{2} = \boxed{\frac{\pi^2}{8}}. \qquad \square$$

2.5   The (Real-Valued) Zeta Function

Although the ζ character was not used before Riemann's Paper, we will follow modern zeta function terminology for our historical look back during the time of Euler.

In its simplest form, the real valued zeta function, $\zeta(x)$, is a function of a real variable that is equal to the sum of the following infinite series:

$$\zeta(x) = \sum_{n=1}^{\infty} \frac{1}{n^x}.$$

Note the following:

- $\zeta(1)$ is the Harmonic Series and $\zeta(2)$ is the Basel Problem.
- The variable x does not need to be a whole number; it can be any real number.
- Recall that for $x = 1$ (the Harmonic Series), the series diverges.
- It is obvious that for any $x < 1$, the series also diverges.

So, at least for real values of the zeta function, the only cases of interest are when $x > 1$.

2.5.1 Convergence

We start by considering the convergence of $\zeta(x)$ for $x > 1$.

Lemma 2.2. *Let $x > 1$, and let $\zeta(x) = \sum\limits_{n=1}^{\infty} \dfrac{1}{n^x}$. Then, $\zeta(x)$ converges.*

Proof. Let $t \geq 1$ and $f(t) = \dfrac{1}{t^x}$ so that $\zeta(x) = \sum\limits_{n=1}^{\infty} f(n)$. Because $x > 1$, $f(t)$ is continuous, positive and decreasing. Thus, by the integral test, $\zeta(x)$ converges if and only if $\int_{1}^{\infty} f(t)\, dt$ converges. We have for $x > 1$:

$$\int_{1}^{a} t^{-x}\, dt = \left(\frac{1}{-x+1} \right) \left(a^{-x+1} - 1 \right) = \left(\frac{-1}{x-1} \right) \left(\frac{1}{a^{x-1}} - 1 \right).$$

Taken to the limit, we have:

$$\int_{1}^{\infty} f(t)\, dt = \lim_{a \to \infty} \int_{1}^{a} t^{-x}\, dt = \lim_{a \to \infty} \left(\frac{-1}{x-1} \right) \left(\frac{1}{a^{x-1}} - 1 \right)$$

$$= \left(\frac{-1}{x-1} \right) \lim_{a \to \infty} \left(\frac{1}{a^{x-1}} - 1 \right) = \left(\frac{-1}{x-1} \right) (0 - 1) = \left(\frac{1}{x-1} \right).$$

But that final fraction is finite for all $x > 1$, completing our proof. $\qquad\square$

2.5.2 The Euler Product Formula

We see that real-valued $\zeta(x)$ converges for $x > 1$ and diverges for $x \leq 1$. At this point, $\zeta(x)$ is interesting but seemingly unrelated to prime numbers. But watch what happens next!

Having dispatched the Basel Problem, Euler turned his attention to $\zeta(x)$. In 1737, Euler derived his ground-breaking *Euler Product Formula*, which became the launching pad for a much deeper study of prime numbers.

Lemma 2.3. *Let $x > 1$. Then, $\zeta(x)$ (an infinite sum of fractions) is equivalent to an infinite product (over the prime numbers), as follows:*

$$\zeta(x) = \sum_{n=1}^{\infty} \frac{1}{n^x} = \prod_{p} \frac{1}{(1 - p^{-x})} \qquad \textit{(where the variable p ranges over the primes).}$$

Proof. See Theorem 2.1, where the equation is proved for $\{s \in \mathbb{C} : Re(s) > 1\}$. $\qquad\square$

With this amazing discovery by Euler, $\zeta(x)$ is no longer just an interesting function. It is now a function that: (1) plays a central role in the study of prime numbers, and (2) launches a whole new area of mathematics: Analytic Number Theory.

By the way, the Euler Product Formula allows for a one-line proof that there are infinitely many primes. Recall that $\zeta(1)$ is the Harmonic Series. Because the Harmonic Series is divergent, that means that the above product of primes (with $x = 1$) is divergent, which is only possible if there are infinitely many primes.

2.6   The (Complex-Valued) Zeta Function

There is really no difficulty in extending $\zeta(s)$ to a function of a complex variable. For $\zeta(s)$, it is traditional to use $s = \sigma + it$ as the complex variable. So we have

$$\zeta(s) = \sum_{n=1}^{\infty} \frac{1}{n^s} \quad s = \sigma + it,\ \sigma > 1, \tag{2.2}$$

with $|n^s| = |e^{\sigma \log|n| + i\theta}| = e^{\sigma \log n} = n^{\sigma}$. Thus, the series converges (and is analytic) for $\sigma > 1$.

We now turn to the complex-valued *Euler Product Formula*.

Theorem 2.1. *Let* $\{s \in \mathbb{C} : Re(s) > 1\}$. $\zeta(s)$ *(an infinite sum of fractions) is equivalent to an infinite product (over the prime numbers), as follows:*

$$\zeta(s) = \sum_{n=1}^{\infty} \frac{1}{n^s} = \prod_{p} \frac{1}{(1 - p^{-s})} \quad \text{(where the variable p ranges over the primes).}$$

Proof #1 – Classic Sieve. For convergence of the product, let p_n be the nth prime and $s = \sigma + it$:

$$\sum_{n=1}^{\infty} \left| \frac{1}{p_n^s} \right| = \sum_{n=1}^{\infty} \frac{1}{p_n^{\sigma}} \leq \sum_{n=1}^{\infty} \frac{1}{n^{\sigma}}.$$

Because the last sum converges for $\sigma > 1$, our product converges by lemma 11.2.

To prove our equality, we start with $\zeta(s)$ and multiply by the first fractional term on the RHS:

$$\zeta(s) = 1 + \frac{1}{2^s} + \frac{1}{3^s} + \frac{1}{4^s} + \frac{1}{5^s} \cdots$$

$$\frac{1}{2^s} \zeta(s) = \frac{1}{2^s} \left(1 + \frac{1}{2^s} + \frac{1}{3^s} + \frac{1}{4^s} + \frac{1}{5^s} \cdots \right) = \left(\frac{1}{2^s} + \frac{1}{4^s} + \frac{1}{6^s} + \frac{1}{8^s} + \frac{1}{10^s} \cdots \right).$$

Note that the RHS has *only* terms (and *all* such term) with a factor of 2 in the denominator. Now subtract both sides from $\zeta(s)$:

$$\zeta(s) - \frac{1}{2^s} \zeta(s) = \left(1 + \frac{1}{2^s} + \frac{1}{3^s} + \frac{1}{4^s} + \frac{1}{5^s} \cdots \right) - \left(\frac{1}{2^s} + \frac{1}{4^s} + \frac{1}{6^s} + \frac{1}{8^s} + \frac{1}{10^s} \cdots \right)$$

$$\left(1 - \frac{1}{2^s} \right) \zeta(s) = \left(1 + \frac{1}{3^s} + \frac{1}{5^s} + \frac{1}{7^s} + \frac{1}{9^s} \cdots \right). \tag{2.3}$$

We now have a series on the RHS where no term includes a factor of 2 in its denominator. Using the same technique, we multiply both sides by the first fractional term on the RHS:

$$\left(\frac{1}{3^s} \right) \left(1 - \frac{1}{2^s} \right) \zeta(s) = \left(\frac{1}{3^s} \right) \left(1 + \frac{1}{3^s} + \frac{1}{5^s} + \frac{1}{7^s} + \frac{1}{9^s} \cdots \right) = \left(\frac{1}{3^s} + \frac{1}{9^s} + \frac{1}{15^s} + \frac{1}{21^s} \cdots \right). \tag{2.4}$$

Now the RHS has *only* terms with a factor of 3 in the denominator, and *all* such terms **except** terms that include a prime number smaller than 3. We next subtract equation (2.4) from equation (2.3):

$$\left(1 - \frac{1}{2^s} \right) \zeta(s) - \left(\frac{1}{3^s} \right) \left(1 - \frac{1}{2^s} \right) \zeta(s) = \left(1 + \frac{1}{3^s} + \frac{1}{5^s} + \frac{1}{7^s} + \frac{1}{9^s} \cdots \right) - \left(\frac{1}{3^s} + \frac{1}{9^s} + \frac{1}{15^s} + \frac{1}{21^s} \cdots \right)$$

$$\left(1 - \frac{1}{3^s} \right) \left(1 - \frac{1}{2^s} \right) \zeta(s) = \left(1 + \frac{1}{5^s} + \frac{1}{7^s} + \frac{1}{11^s} + \frac{1}{13^s} \cdots \right).$$

We now have a series on the RHS where no term includes a factor of 2 or 3 in its denominator. The first fractional term will have a denominator of 5, the next prime number – it could not have been previously eliminated and all non-prime numbers below 5 must be a factor of one of the previously eliminated prime numbers. As we continue like this, our equation will look like:

$$\left(1 - \frac{1}{p_n^s}\right)\left(1 - \frac{1}{p_{n-1}^s}\right) \cdots \left(1 - \frac{1}{2^s}\right) \zeta(s) = \left(1 + \frac{1}{p_{n+1}^s} + \text{more terms...}\right)$$

where none of the term on the right will include a factor of $p_1, p_2, ...p_n$ in the denominator.

Note the right side is being progressively "sieved". All terms having the first n prime numbers as factors in their denominator are removed. Repeating infinitely with all prime numbers we get:

$$\cdots \left(1 - \frac{1}{13^s}\right)\left(1 - \frac{1}{11^s}\right)\left(1 - \frac{1}{7^s}\right)\left(1 - \frac{1}{5^s}\right)\left(1 - \frac{1}{3^s}\right)\left(1 - \frac{1}{2^s}\right) \zeta(s) = 1$$

Dividing both sides by everything on the left but $\zeta(s)$ we get:

$$\zeta(s) = \frac{1}{\left(1 - \frac{1}{2^s}\right)\left(1 - \frac{1}{3^s}\right)\left(1 - \frac{1}{5^s}\right)\left(1 - \frac{1}{5^s}\right)\left(1 - \frac{1}{11^s}\right)\cdots} = \prod_{p=primes} \frac{1}{\left(1 - p^{-s}\right)}. \qquad \square$$

Proof #2 – Fundamental Theorem of Arithmetic. Because $Re(s) > 1$, we have $p^{-s} < 1$. Thus, each term of our product can be expanded into a geometric series:

$$\frac{1}{1 - p^{-s}} = 1 + \frac{1}{p^s} + \frac{1}{p^{2s}} + \frac{1}{p^{3s}} + \cdots + \frac{1}{p^{ks}} + \cdots$$

Now fix $N \in \mathbb{N}$ and let $s = \sigma + it$. We claim

$$\left| \zeta(s) - \prod_{p \leq N} \frac{1}{\left(1 - p^{-s}\right)} \right| < \sum_{n=N+1}^{\infty} \frac{1}{n^\sigma}.$$

Consider the finite number of terms in the product. Replace each of them by their geometric series and multiply all terms. By the fundamental theorem of arithmetic, the result is the reciprocal of *every* natural number that is *only* factored by primes less than or equal to N. This includes (but is not limited to) all natural numbers $\leq N$, proving the inequality. As $n \to \infty$, the difference between $\zeta(s)$ and the partial product goes to zero, as required to prove the theorem. $\qquad \square$

2.7   The Reciprocal of the Primes

We briefly depart from $\zeta(s)$ to discuss a related and interesting variation. As shown above, $\zeta(1)$ (the Harmonic Series) diverges, but ever so slowly. We visit Euler's work (yet again) as he considers another infinite series: the sum of the reciprocals of the primes. This series is a relatively small subset of $\zeta(1)$, since it includes only the terms of $\zeta(1)$ where the denominator is a prime number. Surely this series converges. Right? Wrong! Euler's proof was published in 1737 (slightly cleaned up in 1741).

Lemma 2.4. *Let $\{p_n\}$ be an ordered set of all primes. Then, the following series diverges:*

$$\sum_{n=1}^{\infty} \frac{1}{p_n}. \tag{2.5}$$

Proof #1 – Euler's Proof. In his proof below, Euler uses the Taylor Series for $\log(x + 1)$ and the formula for the Geometric Series.

$$\ln\left(\sum_{n=1}^{\infty} \frac{1}{n}\right) = \ln\left(\prod_{p} \frac{1}{1 - p^{-1}}\right) = \sum_{p} \ln\left(\frac{1}{1 - p^{-1}}\right) = \sum_{p}\left[\ln(1) - \ln\left(1 - \frac{1}{p}\right)\right]$$

$$= -\sum_{p} \ln\left(1 - \frac{1}{p}\right) = -\sum_{p}\left(-\sum_{n=1}^{\infty} \frac{1}{np^n}\right) = \sum_{p}\left(\sum_{n=1}^{\infty} \frac{1}{np^n}\right)$$

$$= \sum_{p}\left(\frac{1}{p} + \frac{1}{2p^2} + \frac{1}{3p^3} + \ldots\right)$$

$$= \sum_{p} \frac{1}{p} + \sum_{p} \frac{1}{p^2}\left(\frac{1}{2} + \frac{1}{3p} + \frac{1}{4p^2} + \ldots\right)$$

$$< \sum_{p} \frac{1}{p} + \sum_{p} \frac{1}{p^2}\left(1 + \frac{1}{p} + \frac{1}{p^2} + \ldots\right) = \sum_{p} \frac{1}{p} + \sum_{p} \frac{1}{p^2}\left(\frac{1}{1 - p^{-1}}\right)$$

$$= \sum_{p} \frac{1}{p} + \sum_{p}\left(\frac{1}{p(p - 1)}\right)$$

$$\leq \sum_{p} \frac{1}{p} + \sum_{p} \frac{1}{p^2} \leq \sum_{p} \frac{1}{p} + \sum_{n=1}^{\infty} \frac{1}{n^2} = \sum_{p} \frac{1}{p} + \frac{\pi^2}{6}$$

We know that the first x terms of the Harmonic Series sum to approximately $\log(x)$. The above proof by Euler: (1) shows that the sum of the reciprocals of the primes diverges, and (2) shows that the first x terms of the "Prime Harmonic Series" sum to approximately $\log(\log(x))$. $\qquad\square$

Proof #2 – Clarkson. [3] Assume equation (2.5) converges and fix k such that:

$$\sum_{n=k+1}^{\infty} \frac{1}{p_n} < \frac{1}{2}.$$

Let $Q = p_1 \cdot p_2 \cdot \ldots \cdot p_k$ and consider the number $(1 + nQ)$ for $n \in \mathbb{N}$. Either $n = 1$ or n consists of one or more prime factors. Thus, nQ will have all of the prime factors of Q (some perhaps multiple times) and possibly some additional prime factors $p_i > p_k$. Therefore, $(1 + nQ)$ is not divisible by any of the first k primes. So the prime factors of $(1 + nQ)$, for all n, must occur among the primes $p_{k+1}, p_{k+2}\cdots$. Thus, for any r:

$$\sum_{n=1}^{r} \frac{1}{(1 + nQ)} \leq \sum_{t=1}^{\infty}\left(\sum_{n=k+1}^{\infty}\left(\frac{1}{p_n}\right)\right)^t < \sum_{t=1}^{\infty}\left(\frac{1}{2}\right)^t = 1.$$

The key is to note that every fraction in the left sum (consisting only of prime factors $> p_k$) can be found in one of the expansions of $(1/p_{k+1} + 1/p_{k+2} + 1/p_{k+3} + \ldots)^t$. But that means, taken to the limit, the left sum converges. And that leads to our contradiction because, as the following equation shows, that would imply that the Harmonic Series converges:

$$\sum_{n=1}^{\infty} \frac{1}{(1 + nQ)} \geq \sum_{n=1}^{\infty} \frac{1}{n(1 + Q)} = \frac{1}{(1 + Q)} \sum_{n=1}^{\infty} \frac{1}{n}. \qquad\square$$

Proof #3 – Hardy and Wright. [4] We start by describing certain properties of square-free numbers.

> A positive integer is **square-free** if its decomposition contains no repeated factors. Clearly, all prime numbers are square-free. By convention, the number 1 is square-free. All of the square-free numbers that can be built from the first p_j primes look like:
>
> $$p_1^{a_1} \cdot p_2^{a_2} \cdot p_3^{a_3} \cdot \ldots \cdot p_j^{a_j},$$
>
> where each a is either 0 or 1. That means there are 2^j square-free numbers that can be built from the primes up to p_j. (If all a are 0, we have the square-free number 1).
>
> We describe one more property of square-free numbers. Every positive integer N can be built from $N = (m^2)k$, where m and k are positive integers and k is a square-free number. If N is already a square-free integer, then $m = 1$. Otherwise, for each prime factor of N having a power greater than one, set m^2 equal to the product of the highest available even power of each such prime factor. For example, if $N = 2^2 \cdot 3 \cdot 5 \cdot 7^3$, then $m^2 = 2^2 \cdot 7^2$ and $k = 3 \cdot 5 \cdot 7$.

For a given prime p_i and a given positive whole number N, we define two functions:

> $\mathbf{N_{Sm}(p_i, N)}$ is equal to the number of positive integers less than or equal to N which have only "small" prime factors (i.e., no prime factor that is greater than p_i). For n included in $N_{Sm}(p_i, N)$, we can write n using the $n = (m^2)k$ form described above. We saw above, there can be only 2^i square-free parts to $n = (m^2)k$. Because $m \le \sqrt{n} \le \sqrt{N}$, there can be only $\sqrt{N}$ square parts to $n = (m^2)k$. Thus we have:
>
> $$N_{Sm}(p_i, N) \le 2^i \sqrt{N}.$$

> $\mathbf{N_{Bg}(p_i, N)}$ is equal to the number of positive integers less than or equal to N which have at least one "big" prime factor (i.e., a prime factor greater than p_i). Note that the number of positive integers $n \le N$ that are divisible by a given prime p is $\lfloor N/p \rfloor$ where $\lfloor x \rfloor$ is the floor function (i.e., greatest integer $\le x$). Thus, we have:
>
> $$N_{Bg}(p_i, N) \le \sum_{n=i+1}^{\infty} \left\lfloor \frac{N}{p_n} \right\rfloor.$$

By definition, we have $N_{Sm}(p_i, N) + N_{Bg}(p_i, N) = N$.

Now assume equation (2.5) converges. Then we can fix i such that for primes greater than p_i we have (the vanishing remainder of the series):

$$\sum_{n=i+1}^{\infty} \frac{1}{p_n} < \frac{1}{2} \quad \text{which means} \quad N_{Bg}(p_i, N) \le \sum_{n=i+1}^{\infty} \left\lfloor \frac{N}{p_n} \right\rfloor < \frac{N}{2}.$$

Now set $N = 2^{2i+2}$. Using our inequalities, we have:

$$N_{Sm}(p_i, N) = N_{Sm}(p_i, 2^{2i+2}) \le 2^i \sqrt{2^{2i+2}} = 2^i \cdot 2^{i+1} = 2^{2i+1}$$

$$N_{Bg}(p_i, N) = N_{Bg}(p_i, 2^{2i+2}) < \frac{2^{2i+2}}{2} = 2^{2i+1}$$

Our convergence assumption leads to a contradiction because $N_{Sm}(p_i, N) + N_{Bg}(p_i, N) < N$. □

Chapter 3

The Factorial (to the Gamma Function)

3.1 Introduction

Once again, Leonhard Euler is the leading player in a very important advance in mathematics, the development of the gamma function. We will later see that the zeta function needs the assistance of the gamma function before it can open the door to the secrets of prime numbers.

We start with a simple concept: the factorial function, which applies to real, non-negative whole numbers. Specifically:

$$n! = 1 \cdot 2 \cdot 3 \cdot ... \cdot n = \prod_{k=1}^{n} k.$$

So, $n!$ is the product of all whole numbers between 1 and the positive whole number n. The expression $0!$ is treated as a special cases and is defined to equal 1.

Can the factorial function be extended to apply when n is not a whole number? We pick up the story from an article by Philip Davis[5] describing the history of the gamma function:

> The year 1729 saw the birth of the gamma function in a correspondence between a Swiss mathematician in St. Petersburg and a German mathematician in Moscow. The former: Leonhard Euler (1707-1783), then 22 years of age, but to become a prodigious mathematician, the greatest of the 18th century. The latter: Christian Goldbach (1690-1764), a savant, a man of many talents and in correspondence with the leading thinkers of the day. As a mathematician, he was something of a dilettante, yet he was a man who bequeathed to the future a problem in the theory of numbers so easy to state and so difficult to prove that even to this day it remains on the mathematical horizon as a challenge [Goldbach's Conjecture].
>
> The birth of the gamma function was due to the merging of several mathematical streams. The first was that of interpolation theory, a very practical subject largely the product of English mathematicians of the 17th century but which all mathematicians enjoyed dipping into from time to time. The second stream was that of the integral calculus and of the systematic building up of the formulas of indefinite integration, a process which had been going on steadily for many years. A certain ostensibly simple problem of interpolation arose and was bandied about unsuccessfully by Goldbach and by Daniel Bernoulli (1700-1784) and even earlier by James Stirling (1692-1770). The problem was posed to Euler. Euler announced his solution to Goldbach in two letters which were to be the beginning of an extensive correspondence which lasted the duration of Goldbach's life. The first letter dated October 13, 1729 dealt with the interpolation problem, while the second dated January 8, 1730 dealt with integration and tied the two together.

Of course, the interpolation problem Euler was addressing was interpolation allowing solutions for non-integer factorial. For example, find a solution for 4.32! More generally, Euler was looking for an analytic function which would yield the expected factorial values when a positive integer was inserted, but which would still be meaningful for other (non-integer) values of the variable.

You may call it experimentation, intuition, instinct, or some similar term, but by whatever means Euler noticed that the following infinite product correctly solves for factorials of positive integers:

$$\left[\left(\frac{2}{1}\right)^n \left(\frac{1}{n+1}\right)\right]\left[\left(\frac{3}{2}\right)^n \left(\frac{2}{n+2}\right)\right]\left[\left(\frac{4}{3}\right)^n \left(\frac{3}{n+3}\right)\right]\ldots = n!.$$

Below is how we would write this today (we call this the **Euler Product**):

$$\lim_{m\to\infty}\left(\frac{m!\,(m+1)^n}{(n+1)\,(n+2)\,(n+3)\,\ldots\,(n+m)}\right) = n!. \tag{3.1}$$

The equation is valid for all $n \in \mathbb{R}$ (other than negative integers) and in fact, solves the interpolation problem. But Euler did not stop there. He noticed that with $n = \frac{1}{2}$ the formula yields (with some manipulation) the famous infinite product of the Englishman John Wallis (1616-1703):

$$\left(\frac{2 \cdot 2}{1 \cdot 3}\right)\left(\frac{4 \cdot 4}{3 \cdot 5}\right)\left(\frac{6 \cdot 6}{5 \cdot 7}\right)\left(\frac{8 \cdot 8}{7 \cdot 9}\right)\ldots = \frac{\pi}{2}.$$

But π and circles made him think of integrals. How could he express his infinite product as an integral? Davis [11, p. 853] provides fascinating details regarding Euler's thought process as he develops the integral representation of the interpolated factorial function. Euler's integral:

$$n! = \int_0^1 (-\log y)^n \, dy \quad \text{(or, equally):} \quad = \int_0^1 (\log 1/y)^n \, dy.$$

The modern form of the integral is obtained by change of variable $x = -\log(y)$ so that $y = e^{-x}$ and $dx = -(dy/y)$ and $-e^{-x}dx = dy$. This also changes the limits of integration, with: $y = 0 \to x = \infty$ and $y = 1 \to x = 0$. We have:

$$n! = -\int_\infty^0 e^{-x}x^n \, dx = \int_0^\infty e^{-x}x^n \, dx.$$

In 1813, Gauss introduced the Π (Pi) function:

$$\Pi(s) = s! = \int_0^\infty e^{-x}x^s \, dx \quad (s \in \mathbb{C}, Re(s) > -1),$$

At about the same time, Legendre introduced the Γ (Gamma) function:

$$\Gamma(s) = (s-1)! = \int_0^\infty e^{-x}x^{s-1} \, dx \quad (s \in \mathbb{C}, Re(s) > 0). \tag{3.2}$$

To the dismay of some, by the end of the 19th century, Legendre's "off by one" **gamma function** prevailed and became the standard integral for expressing the interpolated factorial function.

In this chapter (for $Re(s) > 0$), we will: (1) show that the gamma function integral converges, (2) provide a functional equation based on the gamma function integral, and (3) show that the gamma function integral is equal to the Euler Product. We will then introduce the Weierstrass Form of the gamma function. Next, we provide three proofs of the analytic continuation of the gamma function to the entire complex plane (with simple poles at the non-positive integers). Finally, we compute some specific values of the gamma function – most importantly, we show that $\Gamma(1/2) = \sqrt{\pi}$.

3.2 Convergence of the Gamma Function

Lemma 3.1. *Let $\{s \in \mathbb{C} : Re(s) > 0\}$. In that half-plane, the Γ function (the integral below) extends to an analytic function:*

$$\Gamma(s) = \int_0^\infty e^{-x} x^{s-1} \, dx \quad \{s \in \mathbb{C} : Re(s) > 0\}.$$

Proof.[6] Fix s and let $u = Re(s)$. We have $\left| e^{-x} x^{s-1} \right| = e^{-x} x^{u-1}$. But that means we can test convergence on the real integral. Our potential problems are at the limits of integration. We divide the integral into two parts, testing the lower limits with the first integral and the upper limits with the second.

$$\Gamma(s) = \int_0^1 e^{-x} x^{u-1} \, dx + \int_1^\infty e^{-x} x^{u-1} \, dx.$$

We start with the first integral, where the only potential problem is the lower limit. Near 0, e^{-x} is bounded, with its largest value of 1 when $x = 0$. Thus:

$$\int_0^1 e^{-x} x^{u-1} \, dx \le \int_0^1 1 \cdot x^{u-1} \, dx = \int_0^1 x^{u-1} \, dx = \left. \frac{x^u}{u} \right|_0^1 = \frac{1}{u},$$

which is clearly finite and well-behaved for all $u = Re(s) > 0$.

How about the upper limit of the right integral? Fix $M = u + 2 + \epsilon$ where $0 \le \epsilon \le 1$ and ϵ is chosen to make M an integer. We have $e^x > x^M/M!$ because this is just one term of the Taylor Series expansion for e^x. But that means (taking the reciprocals) that $e^{-x} < M!/x^M$ and we have:

$$\int_1^B e^{-x} x^{u-1} \, dx \le \int_1^B M! x^{-M} x^{u-1} \, dx = M! \int_1^B x^{u-M-1} \, dx = M! \left. \frac{x^{u-M}}{(u-M)} \right|_1^B = \frac{M!}{(u-M)} \left[B^{u-M} - 1 \right]$$

$$= \frac{M!}{(u-M)} \left[\frac{1}{B^{M-u}} - 1 \right] = \frac{M!}{(M-u)} \left[1 - \frac{1}{B^{M-u}} \right] = \frac{M!}{(2+\epsilon)} \left[1 - \frac{1}{B^{2+\epsilon}} \right].$$

Thus

$$\lim_{B \to \infty} \int_1^B e^{-x} x^{u-1} \, dx \le \lim_{B \to \infty} \frac{M!}{(2+\epsilon)} \left[1 - \frac{1}{B^{2+\epsilon}} \right] \le \frac{M!}{(2+\epsilon)} \le \frac{(u+3)!}{2},$$

and the upper limit is clearly finite and well-behaved for all $u = Re(s) > 0$. $\qquad \square$

3.3 A Functional Equation for the Gamma Function

Lemma 3.2. *For $\{s \in \mathbb{C} : Re(s) > 0\}$, $\Gamma(s+1) = s\Gamma(s)$.*

Proof. By definition:

$$\Gamma(s+1) = \int_0^\infty e^{-x} x^{s+1-1} \, dx = \int_0^\infty e^{-x} x^s \, dx.$$

Let $u(x) = x^s$ and $v(x) = -e^{-x}$. By the chain rule: $v'(x) = e^{-x}$. Now, integrate by parts:

$$\int_a^b u(x) v'(x) \, dx = u(x) v(x) \big|_a^b - \int_a^b u'(x) v(x) \, dx,$$

so that

$$\Gamma(s+1) = \int_0^\infty x^s e^{-x} \, dx = -x^s e^{-x} \big|_0^\infty + s \int_0^\infty x^{s-1} e^{-x} \, dx$$

$$= 0 + s \int_0^\infty x^{s-1} e^{-x} \, dx = s\Gamma(s). \qquad \square$$

3.4   Compare the Gamma Function to Euler's Product

In this section, we show that the gamma function integral is equal to the Euler Product in the right half-plane. We start with a useful lemma.

Lemma 3.3. *For $\{s \in \mathbb{C} : Re(s) > 0\}$ and $n \in \mathbb{N}$, we have:*

$$I(s)_n = \int_0^1 x^{s-1}(1-x)^n \, dx = \frac{n!}{s(s+1)(s+2)...(s+n)}.$$

Proof. We integrate by parts. Set $v(x) = x^s/s$ so that $v'(x) = x^{s-1}$. Set $u(x) = (1-x)^n$ so that (using the chain rule) $u'(x) = -n(1-x)^{n-1}$. We have:

$$\int_0^1 x^{s-1}(1-x)^n \, dx = \left[\frac{x^s}{s}(1-x)^n\right]\Big|_0^1 + \int_0^1 \frac{x^s}{s}n(1-x)^{n-1} \, dx = \frac{n}{s}\int_0^1 x^s(1-x)^{n-1} \, dx.$$

We repeat the same process on the new integral so that:

$$\frac{n}{s}\int_0^1 x^s(1-x)^{n-1} \, dx = \frac{n}{s}\frac{(n-1)}{(s+1)}\int_0^1 x^{s+1}(1-x)^{n-2} \, dx = \frac{n}{s}\frac{(n-1)}{(s+1)}\frac{(n-2)}{(s+2)}\int_0^1 x^{s+2}(1-x)^{n-3} \, dx.$$

Now continue the same process n times, until we have:

$$\frac{n!}{s(s+1)(s+2)(s+n-1)}\int_0^1 x^{s+n-1}(1-x)^0 \, dx = \frac{n!}{s(s+1)(s+2)(s+n-1)}\left[\frac{x^{s+n}}{(s+n)}\right]\Big|_0^1,$$

yielding our final result:

$$I(s)_n = \int_0^1 x^{s-1}(1-x)^n \, dx = \frac{n!}{s(s+1)(s+2)...(s+n)}. \qquad \square$$

We can now compare the gamma function integral and Euler's Product.

Lemma 3.4. *For $\{s \in \mathbb{C} : Re(s) > 0\}$ and $n \in \mathbb{N}$:*

$$\Gamma(s) = \int_0^\infty e^{-x}x^{s-1} \, dx = \lim_{n\to\infty} n^s \frac{n!}{s(s+1)(s+2)...(s+n)}.$$

Proof. We use the following known property: $e^{-t} = \lim_{n\to\infty}\left(1 - \frac{t}{n}\right)^n$. Thus, for $Re(s) > 0$, we have (using Lebesgue's Dominated Convergence Theorem to justify moving the limit outside integral):

$$\Gamma(s) = \int_0^\infty e^{-t}t^{s-1} \, dt = \lim_{n\to\infty} \int_0^n \left(1 - \frac{t}{n}\right)^n t^{s-1} \, dt.$$

Now substitute $t = nx$, so that $dt = ndx$, and as the upper limit $t \to n$, the upper limit $x \to n/n = 1$:

$$\Gamma(s) = \lim_{n\to\infty} \int_0^1 n^{s-1}x^{s-1}(1-x)^n n \, dx = \lim_{n\to\infty} n^s \int_0^1 x^{s-1}(1-x)^n \, dx.$$

Using Lemma 3.3:

$$\Gamma(s) = \lim_{n\to\infty} n^s \int_0^1 x^{s-1}(1-x)^n \, dx = \lim_{n\to\infty} n^s I(s)_n = \lim_{n\to\infty} n^s \frac{n!}{s(s+1)(s+2)...(s+n)}. \qquad \square$$

3.5 The Weierstrass Form of the Gamma Function

3.5.1 Definitions

We define

$$S_n = \sum_{k=1}^{n} \frac{1}{k} \quad \text{and} \quad \gamma_n = (S_n - \log n) \quad \text{which means that} \quad S_n = (\gamma_n + \log n).$$

Note that the *Euler-Mascheroni constant* (approximately 0.57722) is defined as

$$\gamma = \lim_{n \to \infty} (S_n - \log n).$$

3.5.2 The Euler-Mascheroni Series Has a Finite Limit

To verify that γ has a finite limit, we first evaluate the following integral[7]

$$\int_n^{n+1} \left[\frac{1}{n} - \frac{1}{x}\right] dx = \int_n^{n+1} \frac{1}{n} dx - \left[\int_1^{n+1} \frac{1}{x} dx - \int_1^n \frac{1}{x} dx\right]$$

$$= \frac{1}{n} - [\log(n+1) - \log(n)]$$

so that

$$\gamma_N = \sum_{n=1}^{N} \frac{1}{n} - \log(N) = \frac{1}{N} + \sum_{n=1}^{N-1} \left[\frac{1}{n} - [\log(n+1) - \log(n)]\right]$$

$$= \frac{1}{N} + \sum_{n=1}^{N-1} \left[\int_n^{n+1} \left[\frac{1}{n} - \frac{1}{x}\right] dx\right].$$

Now set the above integrand to the function $f(x) = (1/n - 1/x)$ for $n \le x \le n+1$. If you plot $f(x)$, it has a positive slope and maintains its greatest value at $x = n+1$. Therefore

$$\int_n^{n+1} \left[\frac{1}{n} - \frac{1}{x}\right] dx \le \int_n^{n+1} \left[\frac{1}{n} - \frac{1}{n+1}\right] dx = \int_n^{n+1} \frac{1}{n(n+1)} dx \le \int_n^{n+1} \frac{1}{n^2} dx = \frac{1}{n^2}.$$

That means

$$\gamma_N = \frac{1}{N} + \sum_{n=1}^{N-1} \left[\int_n^{n+1} \left[\frac{1}{n} - \frac{1}{x}\right] dx\right] \le \frac{1}{N} + \sum_{n=1}^{N-1} \frac{1}{n^2}.$$

Thus, $\gamma = \lim_{N \to \infty} \gamma_N$ clearly converges, proving the limit defining γ exists.

3.5.3 Obtain the Weierstrass Form From the Euler Product

We start with

$$\Gamma(s) = \lim_{n \to \infty} n^s \frac{n!}{s(s+1)(s+2)...(s+n)} = \lim_{n \to \infty} \frac{1}{s} n^s \cdot \frac{1 \cdot 2 \cdot 3...n}{(s+1)(s+2)...(s+n)}$$

$$= \lim_{n \to \infty} \frac{1}{s} n^s \cdot \left[\frac{(s+1)(s+2)...(s+n)}{1 \cdot 2 \cdot 3...n}\right]^{-1} = \lim_{n \to \infty} \frac{1}{s} n^s \cdot \prod_{m=1}^{n} \left(1 + \frac{s}{m}\right)^{-1}.$$

Now invert the equation and use $n^{-s} = e^{-s \ln n}$:

$$\frac{1}{\Gamma(s)} = s \lim_{n \to \infty} e^{(-\ln n)s} \prod_{m=1}^{n} \left(1 + \frac{s}{m}\right)$$

Multiply and divide the right side by:

$$e^{sS_n} = \prod_{m=1}^{n} e^{s/m},$$

to obtain

$$\frac{1}{\Gamma(s)} = G(s) = s \lim_{n \to \infty} \left[e^{sS_n}\right] e^{(-\ln n)s} \prod_{m=1}^{n} \left(1 + \frac{s}{m}\right) \left[e^{-s/m}\right]$$

$$= s \lim_{n \to \infty} e^{s\gamma_n} \prod_{m=1}^{n} \left(1 + \frac{s}{m}\right) e^{-s/m} = s e^{s\gamma} \prod_{m=1}^{\infty} \left(1 + \frac{s}{m}\right) e^{-s/m}.$$

The last, which we sometimes refer to as $G(s)$, is the *Weierstrass Form*.

3.6   Analytic Continuance of Gamma Function

We show three different proofs of analytic continuance.

Theorem 3.1. *The gamma function $\Gamma(s)$, initially defined for $Re(s) > 0$, has an analytic continuation to a meromorphic function on $\mathbb{C}$, with simple poles at the non-positive integers. The residue of $\Gamma(s)$ at $s = -n$ is $(-1)^n/n!$.*

Proof #1 – Using Functional Equation.[8]   For $Re(s) > -1$, we define:

$$F_1(s) = \frac{\Gamma(s+1)}{s}.$$

Since $\Gamma(s+1)$ is holomorphic in $Re(s) > -1$, we see that F_1 is meromorphic in that half-plane, with a simple pole at $s = 0$. By the functional equation, if $Re(s) > 0$, we have

$$F_1(s) = \frac{\Gamma(s+1)}{s} = \Gamma(s).$$

Continuing further into the left half of the complex plane, For $Re(s) > -2$, we define:

$$F_2(s) = \frac{\Gamma(s+2)}{s(s+1)}.$$

Since $\Gamma(s+2)$ is holomorphic in $Re(s) > -2$, we see that F_2 is meromorphic in that half-plane, with simple poles at $s = 0$ and $s = -1$. If $Re(s) > 0$, we have by the functional equation

$$F_2(s) = \frac{\Gamma(s+2)}{s(s+1)} = \Gamma(s).$$

We can use the same technique for any negative number $-m$, where for $Re(s) > -m$, we define:

$$F_m(s) = \frac{\Gamma(s+m)}{s(s+1)(s+2)...(s+m-1)}.$$

Since $\Gamma(s+m)$ is holomorphic in $Re(s) > -m$, we see that F_m is meromorphic in that half-plane, with simple poles at $s = 0, -1, -2, ...(-m+1)$. If $Re(s) > 0$, then $F_m(s) = \Gamma(s)$ by the functional equation.

We compute the residues:

$$\begin{aligned} res_{s=-n}F_m(s) &= \frac{\Gamma(-n+m)}{(m-1-n)!(-1)(-2)...(-n)} \\ &= \frac{(m-1-n)!}{(m-1-n)!(-1)(-2)...(-n)} \\ &= \frac{(-1)^n}{n!}. \end{aligned}$$

By the uniqueness theorem, we have $F_m = F_k$ for $1 \le k \le m$ on the domain of definition of F_k. Taking F_m to the limit, we have our analytic continuation of Γ. $\qquad\square$

Proof #2 – Using Split Integrals. [9] For $Re(s) > 0$, we split the Gamma integral:

$$\Gamma(s) = \int_0^1 e^{-x}x^{s-1}\,dx + \int_1^\infty e^{-x}x^{s-1}\,dx. \qquad (3.3)$$

The right integral defines an entire function, so consider only the left integral. Our approach is to expand e^z into a power series and integrate term-by-term:

$$\begin{aligned} \int_0^1 e^{-x}x^{s-1}\,dx &= \int_0^1 \left[\sum_{n=0}^\infty \frac{-x^n}{n!}\right]x^{s-1}\,dx = \int_0^1 \left[\sum_{n=0}^\infty \frac{(-1)^n x^n}{n!}\right]x^{s-1}\,dx \\ &= \sum_{n=0}^\infty \frac{(-1)^n}{n!}\int_0^1 x^{n+s-1}\,dx = \sum_{n=0}^\infty \frac{(-1)^n}{n!}\left[\frac{x^{n+s}}{(n+s)}\Big|_0^1\right] = \sum_{n=0}^\infty \frac{(-1)^n}{n!(n+s)}. \end{aligned}$$

Combining this last result with the right integral in equation (3.3), we have for $Re(s) > 0$:

$$\Gamma(s) = \sum_{n=0}^\infty \frac{(-1)^n}{n!(n+s)} + \int_1^\infty e^{-x}x^{s-1}\,dx. \qquad (3.4)$$

It remains to check whether the series is meromorphic on $\mathbb{C}$. We fix $R > 0$ and split the series into two parts:

$$\sum_{n=0}^\infty \frac{(-1)^n}{n!(n+s)} = \sum_{n=0}^N \frac{(-1)^n}{n!(n+s)} + \sum_{n=N+1}^\infty \frac{(-1)^n}{n!(n+s)},$$

where N is an integer with $N > 2R$. We now consider the disk $|s| < R$. The first (finite) sum defines a meromorphic function in the disk, with the expected poles and residues. For the second sum, we note that $n > N > 2R$ and therefore $|n+s| \ge R$, so that:

$$\left|\frac{(-1)^n}{n!(n+s)}\right| \le \frac{1}{n!R}.$$

Clearly, the second sum is holomorphic in the disk. Since R was arbitrarily chosen, the result holds for all R. Thus, the series in equation (3.4) is meromorphic on $\mathbb{C}$, completing our proof. $\qquad\square$

Proof #3 – Using the Weierstrass Form. [10] We will show the *Weierstrass Form* is holomorphic in $\mathbb{C}$.

Fix $s \in \mathbb{C}$. Note that the factor $se^{s\gamma}$ is fixed, so does not impact convergence. Thus, we consider only

$$\log\left[\prod_{m=1}^{\infty}\left(1 + \frac{s}{m}\right)e^{-s/m}\right] = \sum_{m=1}^{\infty}\left[\log\left(1 + \frac{s}{m}\right) - \frac{s}{m}\right].$$

Using the principal value of $\log s$, $-\pi < \arg(s) \le \pi$, we now study (temporarily disregarding validity of the Taylor Series when $|s/m| > 1$):

$$\left|\log\left(1 + \frac{s}{m}\right) - \frac{s}{m}\right| = \left|\left[\sum_{k=1}^{\infty}\left(\frac{s}{m}\right)^{k}\frac{(-1)^{k-1}}{k}\right] - \frac{s}{m}\right| = \left|\left(\frac{s}{m} - \frac{s}{m}\right) + \left(-\frac{s^2}{2m^2} + \frac{s^3}{3m^3} - \cdots\right)\right|$$

$$= \left|-\frac{s^2}{2m^2} + \frac{s^3}{3m^3} - \cdots\right| \le \frac{|s|^2}{m^2}\left(1 + \frac{|s|}{m} + \frac{|s|^2}{m^2} + \cdots\right)$$

Choose integer M so that $|s| \le M/2$. For $m > M$, we have $|s/m| < 1/2$ and also

$$\left|\log\left(1 + \frac{s}{m}\right) - \frac{s}{m}\right| \le \frac{1}{4}\frac{M^2}{m^2}\left(1 + \frac{1}{2} + \frac{1}{2^2} + \frac{1}{2^3} + \cdots\right) \le \frac{1}{2}\frac{M^2}{m^2}.$$

Therefore

$$\sum_{m=M+1}^{\infty}\left[\log\left(1 + \frac{s}{m}\right) - \frac{s}{m}\right]$$

is an absolutely convergent series of analytical functions for all $s \in \mathbb{C}$, so its exponential

$$\prod_{m=M+1}^{\infty}\left(1 + \frac{s}{m}\right)e^{-s/m} \quad \text{(and therefore)} \quad \prod_{m=1}^{\infty}\left(1 + \frac{s}{m}\right)e^{-s/m} \quad \text{(and therefore)} \quad G(s)$$

are holomorphic for all $s \in \mathbb{C}$, with zeros at the non-positive integers. $\square$

3.6.1 Use the Weierstrass Form to Prove *Euler's Reflection Formula*

We now use $G(s)$ to prove *Euler's Reflection Formula*. In what follows, note that the functional equation gives us $\Gamma(-s) = -\Gamma(1-s)/s$.

$$\frac{-s}{\Gamma(s)\Gamma(1-s)} = \frac{1}{\Gamma(s)\Gamma(-s)} = G(s)G(-s)) = -s^2 e^{\gamma s}e^{-\gamma s}\prod_{m=1}^{\infty}\left(1 + \frac{s}{m}\right)e^{-s/m}\left(1 - \frac{s}{m}\right)e^{s/m}$$

$$= -s^2\prod_{m=1}^{\infty}\left(1 - \frac{s^2}{m^2}\right)$$

$$\frac{1}{\Gamma(s)\Gamma(1-s)} = s\prod_{m=1}^{\infty}\left(1 - \frac{s^2}{m^2}\right) = \frac{1}{\pi}\left[\pi s\prod_{m=1}^{\infty}\left(1 - \frac{s^2}{m^2}\right)\right] = \frac{1}{\pi}\left[\sin(\pi s)\right].$$

The last uses the well-known infinite product representation of the sin function. Finally, we can state *Euler's Reflection Formula*:

$$\Gamma(s)\Gamma(1-s) = \frac{\pi}{\sin(\pi s)}.$$

3.7 Some Computed Values of the Gamma Function

3.7.1 Compute a Few Simple Values

$$\Gamma(1) = 0! = \int_0^\infty e^{-x} x^{1-1}\, dx = \int_0^\infty e^{-x}\, dx = -e^{-x}\Big|_0^\infty = 0 - (-1) = 1.$$

For $\Gamma(2)$, we use integration by parts, with $u(x) = x$ and $v(x) = -e^{-x}$ so that $v'(x) = e^{-x}$:

$$\Gamma(2) = 1! = \int_0^\infty e^{-x} x\, dx = -xe^{-x}\Big|_0^\infty + 1\left[\int_0^\infty e^{-x}\, dx\right] = 0 + 1\,[\Gamma(1)] = 1.$$

For $\Gamma(3)$, we again use integration by parts, this time with $u(x) = x^2$ and $v(x) = -e^{-x}$:

$$\Gamma(3) = 2! = \int_0^\infty e^{-x} x^2\, dx = -x^2 e^{-x}\Big|_0^\infty + 2\left[\int_0^\infty xe^{-x}\, dx\right] = 0 + 2\,[\Gamma(2)] = 2.$$

As you see, after computing $\Gamma(1)$, we are simply applying the gamma functional equation to obtain the value of the next whole number. This makes even clearer that the gamma function agrees (albeit, off by one) with the factorial function for positive whole numbers.

3.7.2 Compute $\Gamma(1/2)$

Lemma 3.5. $\Gamma\left(\frac{1}{2}\right) = \sqrt{\pi}$.

Proof #1 – Using Gaussian Integral. We evaluate a well-known Gaussian integral. Set:

$$G = \int_0^\infty e^{-x^2}\, dx \quad \text{and} \quad G^2 = \left(\int_0^\infty e^{-x^2}\, dx\right)\left(\int_0^\infty e^{-y^2}\, dy\right) = \int_0^\infty \int_0^\infty e^{-(x^2+y^2)}\, dxdy.$$

We change variables to polar coordinates: $x = r\cos\theta$, $y = r\sin\theta$, $r^2 = x^2 + y^2$ and $dxdy = r\,drd\theta$. We are integrating over the upper-right quadrant, with $x, y : 0 \to \infty$. In polar coordinates this becomes $r : 0 \to \infty$ and $\theta : 0 \to \pi/2$.

$$G^2 = \int_0^\infty \int_0^\infty e^{-(x^2+y^2)}\, dxdy = \int_{\theta=0}^{\pi/2} \int_{r=0}^\infty e^{-r^2(\cos^2\theta + \sin^2\theta)} r\,dr\,d\theta = \int_{\theta=0}^{\pi/2} \int_{r=0}^\infty e^{-r^2} r\,dr\,d\theta$$

$$= \int_0^{\pi/2} d\theta \cdot \int_0^\infty r e^{-r^2}\, dr = \frac{\pi}{2} \cdot \int_0^\infty r e^{-r^2}\, dr.$$

For the last integral, we change variable to $u = r^2$ so that $du = 2rdr$ and $rdr = du/2$, giving:

$$G^2 = \frac{\pi}{2} \cdot \int_0^\infty r e^{-r^2}\, dr = \frac{\pi}{2} \cdot \frac{1}{2} \int_0^\infty e^{-u}\, du = \frac{\pi}{2} \cdot \frac{1}{2} \cdot 1 = \frac{\pi}{4}$$

$$G = \frac{\sqrt{\pi}}{2}.$$

Now we can evaluate our gamma integral. We have:

$$\Gamma\left(\frac{1}{2}\right) = \int_0^\infty e^{-x} x^{\left(\frac{1}{2}-1\right)}\, dx = \int_0^\infty e^{-x} x^{\left(-\frac{1}{2}\right)}\, dx = \int_0^\infty e^{-x} \frac{1}{\sqrt{x}}\, dx$$

Now make the change of variable $x = u^2$ so that $dx = 2udu$:

$$\Gamma\left(\frac{1}{2}\right) = \int_0^\infty e^{-x} \frac{1}{\sqrt{x}}\, dx = \int_0^\infty e^{-u^2} \frac{1}{u} 2u\, du = 2\left[\int_0^\infty e^{-u^2}\, du\right] = 2\,[G] = \sqrt{\pi}. \qquad \square$$

Proof #2 – Using Euler's Reflection Formula.

$$\Gamma\left(\frac{1}{2}\right)\Gamma\left(1-\frac{1}{2}\right) = \frac{\pi}{\sin(\pi/2)} \quad \text{so that} \quad \left[\Gamma\left(\frac{1}{2}\right)\right]^2 = \pi \quad \text{and} \quad \Gamma\left(\frac{1}{2}\right) = \sqrt{\pi}. \qquad \square$$

Chapter 4

Riemann's 1859 Paper

4.1 Introduction

We present here Riemann's Paper in its entirety. Our comments/annotations to the paper are shown (between ruled lines) in a bold sans serif font.

4.2 Riemann's Paper

On the Number of Prime Numbers less than a
Given Quantity.
Bernhard Riemann
Monatsberichte der Berliner Akademie, November 1859.
Translation © D. R. Wilkins 1998.[11]

Johann Carl Friedrich Gauss was Riemann's first instructor in mathematics (in 1846) at the University of Göttingen. Gauss was instrumental in recommending that Riemann give up his pursuit of a degree in theology and instead study mathematics. Riemann later studied under Peter Gustav Lejeune Dirichlet at the University of Berlin. In 1859, following the death of Dirichlet (who at that time held Gauss's chair at the University of Göttingen), Riemann was promoted to head the mathematics department at the University of Göttingen. This was Riemann's first paper published after that appointment.

Both Gauss and Dirichlet had an interest in the prime number theorem (and Dirichlet an interest more broadly in number theory). In fact, Gauss may well be the first to have identified (not to say proved) the correct asymptote of $x/\log x$.

I believe that I can best convey my thanks for the honour which the Academy has to some degree conferred on me, through my admission as one of its correspondents, if I speedily make use of the permission thereby received to communicate an investigation into the accumulation of the prime numbers; a topic which perhaps seems not wholly unworthy of such a communication, given the interest which Gauss and Dirichlet have themselves shown in it over a lengthy period.

Until Riemann's paper, what we now know as the zeta function was valid only for complex variables s where $Re(s) > 1$. Riemann's first important result, immediately below, was to extend the zeta function to the entire complex plane, with just one simple pole at 1. As is typical of Riemann, he "sketches" a proof, leaving much to fill in. Our fuller proof of the analytic continuation of the zeta function is presented in Chapter 5.

This is a good place to make a general note about Riemann's paper. He uses the older definition of the Gamma Function, denoted by $\Pi(s)$, versus the now more common $\Gamma(s)$. The difference is reconciled by the formula $\Pi(s) = \Gamma(s+1)$.

For this investigation my point of departure is provided by the observation of *Euler* that the product

$$\prod \frac{1}{1 - \frac{1}{p^s}} = \sum \frac{1}{n^s},$$

if one substitutes for p all prime numbers, and for n all whole numbers. The function of the complex variable s which is represented by these two expressions, wherever they converge, I denote by $\zeta(s)$. Both expressions converge only when the real part of s is greater than 1; at the same time an expression for the function can easily be found which always remains valid. On making use of the equation

$$\int_0^\infty e^{-nx} x^{s-1} dx = \frac{\Pi(s-1)}{n^s}$$

one first sees that

$$\Pi(s-1)\zeta(s) = \int_0^\infty \frac{x^{s-1} dx}{e^x - 1}.$$

If one now considers the integral

$$\int \frac{(-x)^{s-1} dx}{e^x - 1}$$

from $+\infty$ to $-\infty$ taken in a positive sense around a domain which includes the value 0 but no other point of discontinuity of the integrand in its interior, then this is easily seen to be equal to

$$(e^{-\pi s i} - e^{\pi s i}) \int_0^\infty \frac{x^{s-1} dx}{e^x - 1},$$

provided that, in the many-valued function $(-x)^{s-1} = e^{(s-1)\log(-x)}$, the logarithm of $-x$ is determined so as to be real when x is negative. Hence

$$2 \sin \pi s \, \Pi(s-1)\zeta(s) = i \int_\infty^\infty \frac{(-x)^{s-1} dx}{e^x - 1},$$

where the integral has the meaning just specified.

This equation now gives the value of the function $\zeta(s)$ for all complex numbers s and shows that this function is one-valued and finite for all finite values of s with the exception of 1, and also that it is zero if s is equal to a negative even integer.

In this section, Riemann very briefly sketches his first (of two) proofs of the functional equation for the zeta function. This proof relies on contour integration and the residue formula. Our much expanded version of this proof is presented in Chapter 6.

If the real part of s is negative, then, instead of being taken in a positive sense around the specified domain, this integral can also be taken in a negative sense around that domain containing all the remaining complex quantities, since the integral taken though values of infinitely large modulus is then infinitely small. However, in the interior of this domain, the integrand has discontinuities only where x becomes equal to a whole multiple of $\pm 2\pi i$, and the integral is thus equal to the sum of the integrals taken in a negative sense around these values. But the integral around the value $n2\pi i$ is $= (-n2\pi i)^{s-1}(-2\pi i)$, one obtains from this

$$2 \sin \pi s\, \Pi(s-1) \zeta(s) = (2\pi)^s \sum n^{s-1}((-i)^{s-1} + i^{s-1}),$$

thus a relation between $\zeta(s)$ and $\zeta(1-s)$, which, through the use of known properties of the function Π, may be expressed as follows:

$$\Pi\left(\frac{s}{2} - 1\right) \pi^{-\frac{s}{2}} \zeta(s)$$

remains unchanged when s is replaced by $1-s$.

In this section, Riemann very briefly sketches his second (of two) proofs of the functional equation for the zeta function. This proof relies on Fourier transforms, the Poisson Summation Formula, and a Jacobi Theta Function. Our much expanded version of this proof is presented in Chapter 7.

This property of the function induced me to introduce, in place of $\Pi(s-1)$, the integral $\Pi\left(\frac{s}{2} - 1\right)$ into the general term of the series $\sum \frac{1}{n^s}$, whereby one obtains a very convenient expression for the function $\zeta(s)$. In fact

$$\frac{1}{n^s}\Pi\left(\frac{s}{2} - 1\right)\pi^{-\frac{s}{2}} = \int_0^\infty e^{-nn\pi x} x^{\frac{s}{2}-1} dx,$$

thus, if one sets

$$\sum_1^\infty e^{-nn\pi x} = \psi(x)$$

then

$$\Pi\left(\frac{s}{2} - 1\right)\pi^{-\frac{s}{2}}\zeta(s) = \int_0^\infty \psi(x) x^{\frac{s}{2}-1} dx,$$

or since

$$2\psi(x) + 1 = x^{-\frac{1}{2}}\left(2\psi\left(\frac{1}{x}\right) + 1\right), \quad \text{(Jacobi, Fund. S. 184)}$$

$$\Pi\left(\frac{s}{2} - 1\right)\pi^{-\frac{s}{2}}\zeta(s) = \int_1^\infty \psi(x) x^{\frac{s}{2}-1} dx + \int_0^1 \psi\left(\frac{1}{x}\right) x^{\frac{s-3}{2}} dx + \frac{1}{2}\int_0^1 \left(x^{\frac{s-3}{2}} - x^{\frac{s}{2}-1}\right) dx$$

$$= \frac{1}{s(s-1)} + \int_1^\infty \psi(x)\left(x^{\frac{s}{2}-1} + x^{-\frac{1+s}{2}}\right) dx.$$

Riemann's goal in this next section is to show that a certain zeta-related function ($\xi(t)$) can be expanded into an infinite product. Under the right conditions, it can then be treated like a polynomial of infinite degree.

To start, Riemann derives (from the symmetrical form of the functional equation) a new function $\xi(s)$ (sometimes called the "completed zeta function") that: (1) is an entire

function, (2) has zeros at (and only at) the locations of the ζ function's non-trivial zeros, and (3) has the convenient functional equation $\xi(s) = \xi(1-s)$. In our analysis of $\xi(s)$, we will follow Edwards [13, p. 16] (and many others) in rejecting Riemann's change of variable $s = 1/2 + it$ as unnecessarily confusing. So we will consider Riemann's definition below to be of $\xi(s)$ *without* the change of variable to $\xi(t)$. As a nod to Riemann, the Ξ function is defined in the literature to include Riemann's change of variable, so that $\Xi(z) = \xi(1/2 + iz)$. We will have occasion to reference $\Xi(z)$ in our comments below.

Having defined $\xi(s)$, Riemann first restates the function in a wholly different form. This form is the first step in Riemann's development of the infinite product. We provide a much-expanded version of this proof in Chapter 8.

Next, Riemann states (with the barest hint of a proof) an estimating formula for the approximate number of zeros of $\xi(s)$ whose imaginary part lies between 0 and T. It was not until 1905 that von Mangoldt published the first proof showing Riemann's estimating formula was correct. In Chapter 9, we present a proof based on Bäcklund's later (1914 and 1918) improvements to von Mangoldt's original proof.

The following two sentences (a comment by Riemann relating to his estimating formula) are *by far* the most famous in his paper:

> One now finds indeed approximately this number of real roots within these limits, and it is very probable that all roots are real. Certainly one would wish for a stricter proof here; I have meanwhile temporarily put aside the search for this after some fleeting futile attempts, as it appears unnecessary for the next objective of my investigation.

When Riemann says "it is very probable that all roots are real", he is saying it is very probable that the z is real in $\Xi(z) = \xi(1/2 + iz)$ in all cases where $\Xi(z) = 0$. Or, equivalently (as would be expressed today), $\xi(s) = 0$ only if $Re(s) = 1/2$. This is the famous *Riemann Hypothesis*.

The final paragraph is Riemann's conclusion that $\xi(s)$ can be expanded into an infinite product. Actual proof of that proposition requires Hadamard's 1893 proof of the *Hadamard Factorization Theorem*, which itself requires Weierstrass's 1876 proof of the *Weierstrass Factorization Theorem*. We provide proofs of both factorization theorems in Chapter 11. Based on those proofs, we provide a proof that $\xi(s)$ can be expanded into an infinite product in Chapter 8.

I now set $s = \frac{1}{2} + ti$ and

$$\Pi\left(\frac{s}{2}\right)(s-1)\pi^{-\frac{s}{2}}\zeta(s) = \xi(t),$$

so that

$$\xi(t) = \frac{1}{2} - \left(tt + \frac{1}{4}\right)\int_1^\infty \psi(x)x^{-\frac{3}{4}}\cos\left(\frac{1}{2}t\log x\right)dx$$

or, in addition,

$$\xi(t) = 4\int_1^\infty \frac{d(x^{\frac{3}{2}}\psi'(x))}{dx}x^{-\frac{1}{4}}\cos\left(\frac{1}{2}t\log x\right)dx.$$

This function is finite for all finite values of t, and allows itself to be developed in powers of tt as a very rapidly converging series. Since, for a value of s whose real part is greater than 1, $\log \zeta(s) = \sum \log(1 - p^{-s})$ remains finite, and since the same holds for the logarithms of the other factors of $\xi(t)$, it follows that the function $\xi(t)$ can only vanish if the imaginary part of t lies between $\frac{1}{2}i$ and $-\frac{1}{2}i$. The number of roots of $\xi(t) = 0$, whose real parts lie between 0 and T is approximately

$$= \frac{T}{2\pi} \log \frac{T}{2\pi} - \frac{T}{2\pi};$$

because the integral $\int d \log \xi(t)$, taken in a positive sense around the region consisting of the values of t whose imaginary parts lie between $\frac{1}{2}i$ and $-\frac{1}{2}i$ and whose real parts lie between 0 and T, is (up to a fraction of the order of magnitude of the quantity $\frac{1}{T}$) equal to $\left(T \log \frac{T}{2\pi} - T\right) i$; this integral however is equal to the number of roots of $\xi(t) = 0$ lying within in this region, multiplied by $2\pi i$. One now finds indeed approximately this number of real roots within these limits, and it is very probable that all roots are real. Certainly one would wish for a stricter proof here; I have meanwhile temporarily put aside the search for this after some fleeting futile attempts, as it appears unnecessary for the next objective of my investigation.

If one denotes by α all the roots of the equation $\xi(t) = 0$, one can express $\log \xi(t)$ as

$$\sum \log \left(1 - \frac{tt}{\alpha \alpha}\right) + \log \xi(0);$$

for, since the density of the roots of the quantity t grows with t only as $log \frac{t}{2\pi}$, it follows that this expression converges and becomes for an infinite t only infinite as $t \log t$; thus it differs from $\log \xi(t)$ by a function of tt, that for a finite t remains continuous and finite and, when divided by tt, becomes infinitely small for infinite t. This difference is consequently a constant, whose value can be determined through setting $t = 0$.

In this final section of the paper, Riemann develops his "Explicit Formula" for the prime counting function – a difficult and intricate undertaking. Although this is by far the longest section of Riemann's paper, there are still many details left to the reader to work out. In fact, it wasn't until 1895 that Von Mangoldt provided a complete proof (in a different form) of Riemann's explicit formula. Our much expanded version of Riemann's proof is presented in Chapter 10. Von Mangoldt's proof is presented in Chapter 13.

With the assistance of these methods, the number of prime numbers that are smaller than x can now be determined.

Let $F(x)$ be equal to this number when x is not exactly equal to a prime number; but let it be greater by $\frac{1}{2}$ when x is a prime number, so that, for any x at which there is a jump in the value in $F(x)$,

$$F(x) = \frac{F(x+0) + F(x-0)}{2}$$

If in the identity

$$\log \zeta(s) = -\sum \log(1 - p^{-s}) = \sum p^{-s} + \frac{1}{2} \sum p^{-2s} + \frac{1}{3} \sum p^{-3s} + \dots$$

one now replaces

$$p^{-s} \text{ by } s \int_p^\infty x^{-s-1}ds, \quad p^{-2s} \text{ by } s \int_{p^2}^\infty x^{-s-1}ds, ...,$$

one obtains

$$\frac{\log \zeta(s)}{s} = \int_1^\infty f(x)x^{-s-1}dx,$$

if one denotes

$$F(x) + \frac{1}{2}F\left(x^{\frac{1}{2}}\right) + \frac{1}{3}F\left(x^{\frac{1}{3}}\right) + \dots$$

by $f(x)$.

This equation is valid for each complex value $a + bi$ of s for which $a > 1$. If, though, the equation

$$g(s) = \int_0^\infty h(x)x^{-s}d\log x$$

holds within this range, then, by making use of *Fourier's* theorem, one can express the function h in terms of the function g. The equation decomposes, if $h(x)$ is real and

$$g(a + bi) = g_1(b) + ig_2(b),$$

into the two following:

$$g_1(b) = \int_0^\infty h(x)x^{-a}\cos(b\log x)d\log x,$$

$$ig_2(b) = -i\int_0^\infty h(x)x^{-a}\sin(b\log x)d\log x.$$

If one multiplies both equations with

$$(\cos(b\log y) + i\sin(b\log y))db$$

and integrates them from $-\infty$ to $+\infty$ then one obtains $\pi h(y)y^{-a}$ on the right hand side in both, on account of *Fourier's* theorems; thus, if one adds both equations and multiplies them by iy^a, one obtains

$$2\pi ih(y) = \int_{a-\infty i}^{a+\infty i} g(s)y^s ds,$$

where the integration is carried out so that the real part of s remains constant.

For a value of y at which there is a jump in the value of $h(y)$, the integral takes on the mean of the values of the function h on either side of the jump. From the manner in which the function f was defined, we see that it has the same property, and hence in full generality

$$f(y) = \frac{1}{2\pi i}\int_{a-\infty i}^{a+\infty i} \frac{\log \zeta(s)}{s}y^s ds.$$

One can substitute for $\log \zeta$ the expression

$$\frac{s}{2}\log \pi - \log(s-1) - \log \Pi\left(\frac{s}{2}\right) + \sum^\alpha \log\left(1 + \frac{\left(s - \frac{1}{2}\right)^2}{\alpha\alpha}\right) + \log \xi(0)$$

found earlier; however the integrals of the individual terms of this expression do not converge, when extended to infinity, for which reason it is appropriate to convert the previous equation by means of integration by parts into

$$f(x) = -\frac{1}{2\pi i}\frac{1}{\log x}\int_{a-\infty i}^{a+\infty i} \frac{d\frac{\log \zeta(s)}{s}}{ds} x^s ds$$

Since

$$-\log \varPi\left(\frac{s}{2}\right) = \lim\left(\sum_{n=1}^{n=m}\log\left(1+\frac{s}{2n}\right) - \frac{s}{2}\log m\right),$$

for $m = \infty$ and therefore

$$-\frac{d\frac{1}{s}\log \varPi\left(\frac{s}{2}\right)}{ds} = \sum_{1}^{\infty}\frac{d\frac{1}{s}\log\left(1+\frac{s}{2n}\right)}{ds},$$

it then follows that all the terms of the expression for $f(x)$, with the exception of

$$\frac{1}{2\pi i}\frac{1}{\log x}\int_{a-\infty i}^{a+\infty i}\frac{1}{ss}\log \xi(0)x^s ds = \log \xi(0),$$

take the form

$$\pm\frac{1}{2\pi i}\frac{1}{\log x}\int_{a-\infty i}^{a+\infty i}\frac{d\left(\frac{1}{s}\log\left(1-\frac{s}{\beta}\right)\right)}{ds}x^s ds.$$

But now

$$\frac{d\left(\frac{1}{s}\log\left(1-\frac{s}{\beta}\right)\right)}{d\beta} = \frac{1}{(\beta-s)\beta},$$

and, if the real part of s is larger than the real part of β,

$$-\frac{1}{2\pi i}\int_{a-\infty i}^{a+\infty i}\frac{x^s ds}{(\beta-s)\beta} = \frac{x^\beta}{\beta} = \int_{\infty}^{x}x^{\beta-1}dx,$$

or

$$= \int_{0}^{x}x^{\beta-1}dx,$$

depending on whether the real part of β is negative or positive. One has as a result

$$\frac{1}{2\pi i}\frac{1}{\log x}\int_{a-\infty i}^{a+\infty i}\frac{d\left(\frac{1}{s}\log\left(1-\frac{s}{\beta}\right)\right)}{ds}x^s ds.$$
$$= -\frac{1}{2\pi i}\int_{a-\infty i}^{a+\infty i}\frac{1}{s}\log\left(1-\frac{s}{\beta}\right)x^s ds$$
$$= \int_{\infty}^{x}\frac{x^{\beta-1}}{\log x}dx + \text{const}.$$

in the first, and

$$= \int_0^x \frac{x^{\beta-1}}{\log x} dx + \text{ const.}$$

in the second case.

In the first case the constant of integration is determined if one lets the real part of β become infinitely negative; in the second case the integral from 0 to x takes on values separated by $2\pi i$, depending on whether the integration is taken through complex values with positive or negative argument, and becomes infinitely small, for the former path, when the coefficient of i in the value of β becomes infinitely positive, but for the latter, when this coefficient becomes infinitely negative. From this it is seen how on the left hand side $\log\left(1 - \frac{s}{\beta}\right)$ is to be determined in order that the constants of integration disappear.

Through the insertion of these values in the expression for $f(x)$ one obtains

$$f(x) = Li(x) - \sum^\alpha \left(Li\left(x^{\frac{1}{2}+\alpha i}\right) + Li\left(x^{\frac{1}{2}-\alpha i}\right)\right) + \int_x^\infty \frac{1}{x^2-1} \frac{dx}{x \log x} + \log \xi(0),$$

if in $\sum^\alpha$ one substitutes for α all positive roots (or roots having a positive real part) of the equation $\xi(\alpha) = 0$, ordered by their magnitude. It may easily be shown, by means of a more thorough discussion of the function ξ, that with this ordering of terms the value of the series

$$\sum \left(Li\left(x^{\frac{1}{2}+\alpha i}\right) + Li\left(x^{\frac{1}{2}-\alpha i}\right)\right) \log x$$

agrees with the limiting value to which

$$\frac{1}{2\pi i} \int_{a-bi}^{a+bi} \frac{d\frac{1}{s} \sum \log\left(1 + \frac{\left(s-\frac{1}{2}\right)^2}{\alpha\alpha}\right)}{ds} x^s ds$$

converges as the quantity b increases without bound; however when reordered it can take on any arbitrary real value.

From $f(x)$ one obtains $F(x)$ by inversion of the relation

$$f(x) = \sum \frac{1}{n} F\left(x^{\frac{1}{n}}\right),$$

to obtain the equation

$$f(x) = \sum (-1)^\mu \frac{1}{m} f\left(x^{\frac{1}{m}}\right),$$

in which one substitutes for m the series consisting of those natural numbers that are not divisible by any square other than 1, and in which μ denotes the number of prime factors of m.

If one restricts $\sum^\alpha$ to a finite number of terms, then the derivative of the expression for $f(x)$ or, up to a part diminishing very rapidly with growing x,

$$\frac{1}{\log x} - 2 \sum^\alpha \frac{\cos(\alpha \log x) x^{-\frac{1}{2}}}{\log x}$$

gives an approximating expression for the density of the prime number + half the density of the squares of the prime numbers + a third of the density of the cubes of the prime numbers etc. at the magnitude x.

The known approximating expression $F(x) = Li(x)$ is therefore valid up to quantities of the order $x^{\frac{1}{2}}$ and gives somewhat too large a value; because the non-periodic terms in the expression for $F(x)$ are, apart from quantities that do not grow infinite with x:

$$Li(x) - \frac{1}{2}Li\left(x^{\frac{1}{2}}\right) - \frac{1}{3}Li\left(x^{\frac{1}{3}}\right) - \frac{1}{5}Li\left(x^{\frac{1}{5}}\right) + \frac{1}{6}Li\left(x^{\frac{1}{6}}\right) - \frac{1}{7}Li\left(x^{\frac{1}{7}}\right) + \cdots$$

Indeed, in the comparison of $Li(x)$ with the number of prime numbers less than x, undertaken by *Gauss* and *Goldschmidt* and carried through up to $x =$ three million, this number has shown itself out to be, in the first hundred thousand, always less than $Li(x)$; in fact the difference grows, with many fluctuations, gradually with x. But also the increase and decrease in the density of the primes from place to place that is dependent on the periodic terms has already excited attention, without however any law governing this behaviour having been observed. In any future count it would be interesting to keep track of the influence of the individual periodic terms in the expression for the density of the prime numbers. A more regular behaviour than that of $F(x)$ would be exhibited by the function $f(x)$, which already in the first hundred is seen very distinctly to agree on average with $Li(x) + \log \xi(0)$.

Chapter 5

Zeta – Analytic Continuance

5.1 Introduction

In this chapter, we provide a proof of the analytic continuance of $\zeta(s)$ to the full complex plane (with one simple pole at $s = 1$). This proof is a detailed version of the proof sketched in Riemann's Paper. First, we provide two lemmas that will be useful in the proof.

5.2 Two Lemmas

Lemma 5.1. *For $\{z \in \mathbb{C} : |z| \leq \epsilon < 1\}$, $|e^z - 1| \geq |z|(1 - \epsilon)$.*

Proof. We have:

$$\frac{e^z - 1}{z} = \frac{\left(\sum_{k=0}^{\infty} \frac{z^k}{k!}\right) - 1}{z} = \frac{\sum_{k=1}^{\infty} \frac{z^k}{k!}}{z} = \sum_{k=1}^{\infty}\left(\frac{z^{k-1}}{k!}\right) = 1 + \sum_{k=2}^{\infty}\left(\frac{z^{k-1}}{k!}\right).$$

Now fix ϵ and use the triangle inequality:

$$\left|\frac{e^z - 1}{z}\right| = \frac{|e^z - 1|}{|z|} = \left|1 + \sum_{k=2}^{\infty}\left(\frac{z^{k-1}}{k!}\right)\right|$$

$$= \left|1 - \left(-\sum_{k=2}^{\infty}\left(\frac{z^{k-1}}{k!}\right)\right)\right|$$

$$\geq |1| - \left|-\sum_{k=2}^{\infty}\left(\frac{z^{k-1}}{k!}\right)\right| = 1 - \left|\sum_{k=2}^{\infty}\left(\frac{z^{k-1}}{k!}\right)\right|$$

$$\geq 1 - \left|\sum_{k=2}^{\infty} \frac{|z^{k-1}|}{k!}\right| \quad \text{(By induction and repeated use of the triangle inequality)}$$

$$\geq 1 - \left|\sum_{k=2}^{\infty} \frac{\epsilon}{k!}\right| \quad \text{(Because } |z^{k-1}| \leq |z| \leq \epsilon < 1\text{)}$$

$$= 1 - \epsilon \sum_{k=2}^{\infty} \frac{1}{k!} = 1 - \epsilon(e - 2) \geq (1 - \epsilon).$$

Thus: $|e^z - 1| \geq |z|(1 - \epsilon)$. $\qquad\square$

Corollary 5.1. *For $\{z \in \mathbb{C} : |z| \leq 1/2\}$, $|e^z - 1| \geq |z|/2$.*

Proof. This result is immediate from lemma 5.1. $\qquad\square$

Lemma 5.2. *For* $\{z \in \mathbb{C} \setminus [0, \infty) : Re(s) > 1\}$, *we have* $\lim_{\delta \to 0^+} \int_{|z|=\delta} \frac{-z^{s-1}}{e^z - 1} dz = 0$.

Proof. [12] Fix s. Let $A = (Re(s) - 1)$ and let $B = Im(s)$. We use the ML Inequality to obtain an upper bound for our integral. We also assume that $\delta < 1/2$ and apply corollary 5.1:

$$\int_{|z|=\delta} \frac{-z^{s-1}}{e^z - 1} dz \le 2\pi\delta \cdot \max_{|z|=\delta} \left| \frac{-z^{s-1}}{e^z - 1} \right| \le 2\pi\delta \cdot \max_{|z|=\delta} \left| \frac{-z^{s-1}}{\delta/2} \right| = 4\pi \cdot \max_{|z|=\delta} \left| -z^{s-1} \right|.$$

So it only remains to show that

$$\lim_{\delta \to 0^+} \max_{|z|=\delta} \left| -z^{s-1} \right| = 0.$$

Using the principal value of the complex logarithm (for argument θ: $-\pi < \theta \le \pi$), we rewrite $-z = \delta e^{i\theta} = e^{\log(\delta)+i\theta}$, so that

$$\left| -z^{s-1} \right| = \left| e^{(\log(\delta)+i\theta)[A+iB]} \right| = \left| e^{A\log(\delta)-B\theta} e^{i(B\log(\delta)+A\theta)} \right| = \left| e^{A\log(\delta)-B\theta} \right| = \delta^A e^{-B\theta} \le \delta^A e^{\pi|B|}.$$

But $e^{\pi|B|}$ is fixed. Because $A > 0$, we have $\lim_{\delta \to 0^+} \delta^A = 0$, completing the proof. $\square$

5.3 Analytic Continuance of the Zeta Function

Theorem 5.1. *Let* $s \in \mathbb{C}$. *We recall the traditional definition of* $\zeta(s)$:

$$\zeta(s) = \sum_{n=1}^{\infty} \frac{1}{n^s} = \prod_p \frac{1}{(1 - p^{-s})} \quad \text{(where p ranges over an ordered list of all the primes),}$$

convergent for $Re(s) > 1$. *For the contour of integration C (a Hankel contour oriented in the positive direction and more fully described below), we have that:*

$$\zeta(s) = \frac{\Gamma(1-s)}{-2\pi i} \int_C \frac{-z^{s-1}}{e^z - 1} dz \qquad (5.1)$$

extends $\zeta(s)$ *to a meromorphic function on* $\mathbb{C}$, *holomorphic except for a simple pole at $s = 1$.*

Proof. Unless otherwise stated, we assume $Re(s) > 1$. We start with the Gamma function:

$$\Gamma(s) = \int_0^\infty e^{-x} x^{(s-1)} dx.$$

Next, we replace x by nt in the integral (so that $dx = ndt$):

$$\Gamma(s) = \int_0^\infty e^{-nt} (nt)^{(s-1)} n\, dt = \int_0^\infty e^{-nt} n^s t^{(s-1)} dt = n^s \int_0^\infty e^{-nt} t^{(s-1)} dt$$

$$n^{-s} \Gamma(s) = \int_0^\infty e^{-nt} t^{(s-1)} dt$$

$$\frac{1}{n^s} = \frac{1}{\Gamma(s)} \int_0^\infty e^{-nt} t^{(s-1)} dt.$$

Now sum both sides over n:

$$\sum_{n=1}^{\infty} \frac{1}{n^s} = \frac{1}{\Gamma(s)} \sum_{n=1}^{\infty} \int_0^\infty e^{-nt} t^{(s-1)} dt$$

$$\zeta(s) = \frac{1}{\Gamma(s)} \int_0^\infty \sum_{n=1}^\infty e^{-nt} t^{(s-1)} \, dt$$

$$\zeta(s) = \frac{1}{\Gamma(s)} \int_0^\infty t^{(s-1)} \sum_{n=1}^\infty e^{-nt} \, dt. \tag{5.2}$$

Because we are assuming $Re(s) > 1$, the interchange of the sum and integral is justified by the uniform convergence of the sum. Notice that the sum over n in the last integral is just a geometric series with the $n = 0$ term missing. (We are justified in using a geometric series, because $0 < e^{-t} < 1$). So, we have:

$$\sum_{n=1}^\infty e^{-nt} = \sum_{n=0}^\infty e^{-nt} - e^{-0t} = \sum_{n=0}^\infty (e^{-t})^n - 1 = \frac{1}{1 - e^t} - 1 = \frac{e^t}{1 - e^t} = \frac{1}{e^t - 1}.$$

But that means we can restate equation (5.2) (and while we are at it replace dt with the slightly more standard dx), and have:

$$\zeta(s) = \frac{1}{\Gamma(s)} \int_0^\infty x^{(s-1)} \sum_{n=1}^\infty e^{-nx} \, dx = \frac{1}{\Gamma(s)} \int_0^\infty \frac{x^{s-1}}{e^x - 1} \, dx.$$

If we set $I(s) = \int_0^\infty \frac{x^{s-1}}{e^x - 1} \, dx$, then we have: $\zeta(s) = \frac{1}{\Gamma(s)} I(s)$.

Our next goal is to evaluate $I(s)$ in the complex plane and see where that takes us. Riemann starts with a related but not identical integral, which we call $I_0(s)$:

$$I_0(s) = \int_C \frac{-z^{s-1}}{e^z - 1} \, dz.$$

The contour of integration C is described as follows. For small positive δ select two points in the complex plane:

$$p_1 = \delta e^{i\delta} = \delta_x + i\delta_y \quad \text{where} \quad \delta_x = \delta \cos(\delta), \delta_y = \delta \sin(\delta)$$

$$p_2 = \delta e^{-i\delta} = \overline{\delta e^{i\delta}} = \delta_x - i\delta_y$$

Our contour C consists of three piecewise smooth curves $C_1 + C_2 + C_3$ defined as follows:

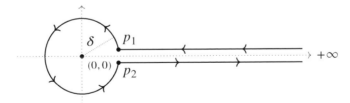

Figure 5.1: Contour of Integration

- C_1 follows a path parallel to (and just above) the real axis from ∞ to p_1.

- C_2 follows a counterclockwise circular path from p_1 to p_2, tracing the boundary of a circle that is centered at 0 with radius δ. (Making just barely less than a full circle).

- C_3 follows a path parallel to (and just below) the real axis from p_2 to ∞.

This contour stays well away from the singularities of the integrand at $2\pi i\mathbb{Z}$. We have:

$$I_0(s) = \int_C \frac{-z^{s-1}}{e^z - 1}\, dz = \int_\infty^{p_1} \frac{-z^{s-1}}{e^z - 1}\, dz + \int_{|z|=\delta} \frac{-z^{s-1}}{e^z - 1}\, dz + \int_{p_2}^\infty \frac{-z^{s-1}}{e^z - 1}\, dz.$$

We now study $I_0(s)$ as $\delta \to 0$. By Lemma 5.2, the second integral can be disregarded so the contour includes only the horizontal lines $\infty \to p_1$ and $p_2 \to \infty$.

Note that $(-z)^{(s-1)} = e^{(s-1)log(-z)}$ and that $log(-z) = log|z| + iarg(-z)$. We will use $Arg(z)$, the principal value of $arg(z)$, defined as the value θ satisfying $-\pi < \theta \le \pi$. Recall that $z \mapsto Arg(z)$ is discontinuous at each point on the nonpositive real axis. Let $z = x_0 + iy$ for some fixed $x_0 < 0$. If $y \downarrow 0$ then $Arg(z) \to \pi$, whereas, if $y \uparrow 0$ then $Arg(z) \to -\pi$. In the first integral, $-z = -\delta_x - i\delta_y$ approaches the nonpositive real axis from below, so (as $\delta \to 0$) we have[13] $Arg(-z) \to -\pi$. In the third integral, $-z = -\delta_x + i\delta_y$ approaches the nonpositive real axis from above, so we have $Arg(-z) \to \pi$.

$$I_0(s) = \int_\infty^{p_1} \frac{e^{(s-1)(\log|z| - \pi i)}}{e^z - 1}\, dz + \int_{p_2}^\infty \frac{e^{(s-1)(\log|z| + \pi i)}}{e^z - 1}\, dz$$

$$= -\int_{p_1}^\infty \frac{e^{(s-1)\log|z|} e^{(s-1)(-\pi i)}}{e^z - 1}\, dz + \int_{p_2}^\infty \frac{e^{(s-1)\log|z|} e^{(s-1)(\pi i)}}{e^z - 1}\, dz$$

$$= -e^{(s-1)(-\pi i)}\int_{p_1}^\infty \frac{e^{(s-1)\log|z|}}{e^z - 1}\, dz + e^{(s-1)(\pi i)}\int_{p_2}^\infty \frac{e^{(s-1)\log|z|}}{e^z - 1}\, dz.$$

Taken to the limit, our integrals are over a horizontal line with $dy = 0$, so we can replace dz with dx (and z with x), replace p_1 and p_2 with 0, and consider $e^{(s-1)\log|z|} = x^{(s-1)}$:

$$\lim_{\delta \to 0+} I_0(s) = -e^{(s-1)(-\pi i)}\int_0^\infty \frac{e^{(s-1)\log|z|}}{e^z - 1}\, dz + e^{(s-1)(\pi i)}\int_0^\infty \frac{e^{(s-1)\log|z|}}{e^z - 1}\, dz$$

$$= \left[e^{i\pi(s-1)} - e^{-i\pi(s-1)}\right]\int_0^\infty \frac{x^{(s-1)}}{e^x - 1}\, dx.$$

But we have: $e^{iw} - e^{-iw} = \cos(w) + i\sin(w) - [\cos(w) - i\sin(w)] = 2i\sin(w)$. So, the value in brackets is: $2i\sin(\pi(s-1)) = 2i\sin(\pi s - \pi) = -2i\sin(\pi s)$. This gives our value for $I_0(s)$:

$$\lim_{\delta \to 0+} I_0(s) = -2i\sin(\pi s)\int_0^\infty \frac{x^{(s-1)}}{e^x - 1}\, dx = -2i\sin(\pi s)I(s).$$

Reviewing, we have:

$$\zeta(s) = \frac{1}{\Gamma(s)}I(s) \quad \text{and} \quad I_0(s) = -2i\sin(\pi s)I(s) \quad \text{so that} \quad \zeta(s) = \frac{1}{\Gamma(s)} \cdot \frac{1}{-2i\sin(\pi s)}I_0(s).$$

Now use a rearrangement of *Euler's Reflection Formula*: $\dfrac{1}{\Gamma(s)} = \dfrac{\Gamma(1-s)\sin(\pi s)}{\pi}$, giving

$$\zeta(s) = \left[\frac{1}{\Gamma(s)}\right]\frac{1}{-2i\sin(\pi s)}I_0(s) = \left[\frac{\Gamma(1-s)\sin(\pi s)}{\pi}\right]\frac{1}{-2i\sin(\pi s)}I_0(s) = \frac{\Gamma(1-s)}{-2\pi i}I_0(s).$$

We therefore have:

$$\zeta(s) = \frac{\Gamma(1-s)}{-2\pi i}\int_C \frac{-z^{s-1}}{e^z - 1}\, dz. \tag{5.3}$$

Up to this point, we have assumed $Re(s) > 1$. But the integral is an entire function (it is uniformly convergent in any compact subset of $\mathbb{C}$ because e^z grows faster than any power of z). That means that $\zeta(s)$ is analytic, except possibly at the positive integers where $\Gamma(1 - s)$ has simple poles. But we already know that $\zeta(s)$ is analytic for $Re(s) > 1$, so the possible poles at $2, 3, 4...$ must be removable and cancel against zeros of the integral. Thus, the only pole of $\zeta(s)$ is at $s = 1$. We discuss in detail the zeros and poles of $\zeta(s)$ in section 7.6. $\qquad\square$

5.4 Analytic Continuance of Zeta Function to $Re(s) > 0$

Above, we have analytically continued $\zeta(s)$ to the full complex plane. However, it is interesting to see how, using elementary methods, we can analytically continue $\zeta(s)$ to the right half plane. In fact, we will find the result useful later in this book. We use here the standard notation that $x = [x] + \{x\}$, where $[x]$ and $\{x\}$ are the integral and fractional parts of the real number x, respectively.

We start with a lemma that applies to the partial summation of certain series[14].

Lemma 5.3 (*Abel's Summation Formula*). *For $n \in \mathbb{N}$, let $\{f_n\}$ be a sequence of complex numbers defined by any arithmetic function $f_n = f(n)$. Define the partial sum $A(x)$ by*

$$A(x) = \sum_{n \leq x} f(n),$$

where $A(x) = 0$ for $x < 1$. Assume the function g has a continuous derivative on the interval $[y, x]$, where $0 < y < x$. Then

$$\sum_{y < n \leq x} f(n)g(n) = A(x)g(x) - A(y)g(y) - \int_y^x A(t)g'(t)\, dt.$$

Proof. Note that $A(x)$ is a step function, with $f(n) = A(n) - A(n - 1)$. Thus

$$\sum_{y < n \leq x} f(n)g(n) = \sum_{n=[y]+1}^{[x]} f(n)g(n) = \sum_{n=[y]+1}^{[x]} [A(n) - A(n - 1)]\, g(n)$$

$$= \sum_{n=[y]+1}^{[x]} A(n)g(n) - \sum_{n=[y]}^{[x]-1} A(n)g(n + 1)$$

$$= \sum_{n=[y]+1}^{[x]-1} A(n)\, [g(n) - g(n + 1)] + A([x])g([x]) - A([y])g([y] + 1)$$

$$= -\sum_{n=[y]+1}^{[x]-1} A(n) \int_n^{n+1} g'(t)\, dt + A([x])g([x]) - A([y])g([y] + 1)$$

$$= -\sum_{n=[y]+1}^{[x]-1} \int_n^{n+1} A(t)g'(t)\, dt + A([x])g([x]) - A([y])g([y] + 1)$$

$$= -\left[\int_{[y]+1}^{[x]} A(t)g'(t)\, dt\right] + \left[A(x)g(x) - \int_{[x]}^x A(t)g'(t)\, dt\right]$$

$$\quad - \left[A(y)g(y) + \int_y^{[y]+1} A(t)g'(t)\, dt\right]$$

$$= A(x)g(x) - A(y)g(y) - \int_y^x A(t)g'(t)\, dt. \qquad\square$$

Lemma 5.4. *The Dirichlet series for* $\zeta(s)$, *holomorphic on* $Re(s) > 1$, *can be analytically continued to the right half plane,* $Re(s) > 0$, *as follows:*

$$\zeta(s) = \sum_{n=1}^{\infty} \frac{1}{n^s} = \frac{s}{s-1} - s \int_1^{\infty} \frac{\{x\}}{x^{s+1}}\, dx, \tag{5.4}$$

with one simple pole at $s = 1$ *having a residue of* 1.

Proof #1. We will use lemma 5.3, setting $y = \frac{1}{2}, f_n = f(n) = 1$ and $g(x) = x^{-s}$. Note that $A(x) = [x]$, $A(y) = 0$, and $g'(x) = -sx^{-s-1}$. Now fix $X \geq 1$. We have

$$
\begin{aligned}
\sum_{n \leq X} n^{-s} &= \frac{[X]}{X^s} - 0 + s \int_{1/2}^{1} \frac{[x]}{x^{s+1}}\, dx + s \int_1^X \frac{[x]}{x^{s+1}}\, dx = s \int_1^X \frac{[x]}{x^{s+1}}\, dx + \frac{[X]}{X^s} \\
&= s \left[\int_1^X \frac{x}{x^{s+1}}\, dx - \int_1^X \frac{x - [x]}{x^{s+1}}\, dx \right] + \left[\frac{X}{X^s} - \frac{X - [X]}{X^s} \right] \\
&= \left[s \int_1^X x^{-s}\, dx - s \int_1^X \frac{x - [x]}{x^{s+1}}\, dx \right] + \left[\frac{1}{X^{s-1}} - \frac{X - [X]}{X^s} \right] \\
&= s \left[\frac{1}{(1-s)x^{s-1}} \Big|_1^X \right] - s \int_1^X \frac{x - [x]}{x^{s+1}}\, dx + \frac{1}{X^{s-1}} - \frac{X - [X]}{X^s} \\
&= \frac{s}{s-1} - \frac{s}{(s-1)X^{s-1}} - s \int_1^X \frac{x - [x]}{x^{s+1}}\, dx + \frac{1}{X^{s-1}} - \frac{X - [X]}{X^s}.
\end{aligned}
\tag{5.5}
$$

As $X \to \infty$, the three fractional terms with X in the denominator go to 0. And $\{x\} = x - [x]$, so that

$$\zeta(s) = \frac{s}{s-1} - s \int_1^{\infty} \frac{\{x\}}{x^{s+1}}\, dx \quad Re(s) > 1.$$

Because $0 \leq \{x\} \leq 1$, the last integral converges and is holomorphic on $Re(s) > 0$. But that means the full equation is meromorphic on $Re(s) > 0$, and thus provides an analytic continuation of $\zeta(s)$ on the half plane $Re(s) > 0$. The $s/(s-1)$ term gives a simple pole at $s = 1$ with residue 1. $\qquad\square$

Proof #2. We assume $Re(s) > 1$ until indicated otherwise.

$$
\begin{aligned}
\zeta(s) = \sum_{n=1}^{\infty} \frac{1}{n^s} &= \sum_{n=1}^{1} \frac{n}{n^s} + \sum_{n=2}^{\infty} \frac{1}{n^s} = \sum_{n=1}^{1} \frac{n}{n^s} + \sum_{n=2}^{\infty} \frac{n - (n-1)}{n^s} \\
&= \sum_{n=1}^{1} \frac{n}{n^s} + \sum_{n=2}^{\infty} \frac{n}{n^s} - \sum_{n=2}^{\infty} \frac{n-1}{n^s} = \sum_{n=1}^{\infty} \frac{n}{n^s} - \sum_{n=2}^{\infty} \frac{n-1}{n^s} \\
&= \sum_{n=1}^{\infty} \frac{n}{n^s} - \sum_{n=1}^{\infty} \frac{n}{(n+1)^s} = \sum_{n=1}^{\infty} n \left[\frac{1}{n^s} - \frac{1}{(n+1)^s} \right] \\
&= s \sum_{n=1}^{\infty} n \int_n^{n+1} x^{-s-1}\, dx.
\end{aligned}
$$

Since $[x] = n$ for any x in the interval $[n, n+1)$, we have

$$
\begin{aligned}
&= s \sum_{n=1}^{\infty} \int_n^{n+1} [x] x^{-s-1}\, dx = s \int_1^{\infty} [x] x^{-s-1}\, dx \\
&= s \left[\int_1^{\infty} x^{-s}\, dx \right] - s \int_1^{\infty} \{x\} x^{-s-1}\, dx \quad (\text{because } [x] = x - \{x\}). \\
&= s \left[\frac{x^{-s+1}}{-s+1} \Big|_1^{\infty} \right] - s \int_1^{\infty} \{x\} x^{-s-1}\, dx,
\end{aligned}
$$

allowing the following simplification

$$\zeta(s) = \frac{s}{s-1} - s \int_1^\infty \frac{\{x\}}{x^{s+1}}\, dx \quad Re(s) > 1.$$

For the same reason as Proof #1, we have our analytic continuation of $\zeta(s)$ on the half plane $Re(s) > 0$, with a simple pole at $s = 1$ with residue 1. $\qquad\square$

5.5 Some Estimates for $\zeta(s)$

Theorem 5.2. *Let $s = \sigma + it \in \mathbb{C}$. We have*

$$|\zeta(s)| < A \log t \qquad\qquad \sigma \geq 1, \quad t \geq 2$$
$$|\zeta(s)| < A(\epsilon) t^{1-\epsilon} \qquad\qquad \sigma \geq \epsilon, \quad t \geq 1, \quad 0 < \epsilon < 1$$

Proof. [15] From Lemma 5.4 and equation (5.5), we have for $\sigma > 0$, $t \geq 1$ and $X \geq 1$:

$$\zeta(s) = \frac{s}{s-1} - s \int_1^\infty \frac{x - [x]}{x^{s+1}}\, dx$$

$$\sum_{n \leq X} n^{-s} = \left[s \int_1^X \frac{x}{x^{s+1}}\, dx - s \int_1^X \frac{x - [x]}{x^{s+1}}\, dx \right] + \left[\frac{X}{X^s} - \frac{X - [X]}{X^s} \right]$$

$$= \left[\frac{s}{s-1} - \frac{s}{(s-1)X^{s-1}} - s \int_1^X \frac{x - [x]}{x^{s+1}}\, dx \right] + \left[\frac{1}{X^{s-1}} - \frac{X - [X]}{X^s} \right].$$

Subtracting the second equality from the first:

$$\zeta(s) - \sum_{n \leq X} n^{-s} = \left[\frac{s}{(s-1)X^{s-1}} - s \int_X^\infty \frac{x - [x]}{x^{s+1}}\, dx \right] - \left[\frac{1}{X^{s-1}} - \frac{X - [X]}{X^s} \right]$$

$$= -s \int_X^\infty \frac{x - [x]}{x^{s+1}}\, dx + \frac{1}{(s-1)X^{s-1}} + \frac{X - [X]}{X^s}$$

Or, equally

$$\zeta(s) = \sum_{n \leq X} n^{-s} - s \int_X^\infty \frac{x - [x]}{x^{s+1}}\, dx + \frac{1}{(s-1)X^{s-1}} + \frac{X - [X]}{X^s}.$$

Therefore

$$|\zeta(s)| \leq \sum_{n \leq X} \frac{1}{n^\sigma} + |s| \int_X^\infty \frac{dx}{x^{\sigma+1}} + \frac{1}{tX^{\sigma-1}} + \frac{1}{X^\sigma}$$

$$\leq \sum_{n \leq X} \frac{1}{n^\sigma} + |s| \left[\frac{x^{-\sigma}}{-\sigma} \Big|_X^\infty \right] + \frac{1}{tX^{\sigma-1}} + \frac{1}{X^\sigma}$$

$$\leq \sum_{n \leq X} \frac{1}{n^\sigma} + \frac{|s|}{\sigma} \frac{1}{X^\sigma} + \frac{1}{tX^{\sigma-1}} + \frac{1}{X^\sigma}$$

$$\leq \sum_{n \leq X} \frac{1}{n^\sigma} + \left(1 + \frac{t}{\sigma}\right) \frac{1}{X^\sigma} + \frac{1}{tX^{\sigma-1}} + \frac{1}{X^\sigma} \quad \text{since } |s| < \sigma + t. \tag{5.6}$$

We now use equation (5.6). For our first claimed inequality, with $\sigma \geq 1$, $t \geq 2$ and $X \geq 1$, we have

$$|\zeta(s)| \leq \sum_{n \leq X} \frac{1}{n} + \frac{1+t}{X} + \frac{1}{t} + \frac{1}{X}$$

$$\leq \sum_{n \leq X} \frac{1}{n} + \frac{t}{X} + \left[\frac{1}{X} + \frac{1}{t} + \frac{1}{X} \right]$$

$$\leq (\log X + 1) + \frac{t}{X} + 3$$

Taking $X = t$, we have our first inequality: $|\zeta(s)| < A \log t$ for $\sigma \geq 1$, $t \geq 2$. ($A = 8$ is sufficient).

We turn to our second claimed inequality, and assume $\sigma \geq \epsilon$, $t \geq 1$ and $0 < \epsilon < 1$.

$$|\zeta(s)| \leq \sum_{n \leq X} \frac{1}{n^\epsilon} + (1 + \frac{t}{\epsilon}) \frac{1}{X^\epsilon} + \frac{1}{tX^{\epsilon-1}} + \frac{1}{X^\epsilon}$$

$$\leq \sum_{n \leq X} \frac{1}{n^\epsilon} + (\frac{2\epsilon}{\epsilon} + \frac{t}{\epsilon}) \frac{1}{X^\epsilon} + \frac{X^{1-\epsilon}}{t}$$

$$\leq \int_0^{[X]} \frac{dx}{x^\epsilon} + \frac{3t}{\epsilon X^\epsilon} + X^{1-\epsilon}$$

$$\leq \frac{X^{1-\epsilon}}{1-\epsilon} + \frac{3t}{\epsilon X^\epsilon} + X^{1-\epsilon}$$

Taking $X = t$, we have our second inequality (with A based on ϵ):

$$|\zeta(s)| \leq \left(\frac{1}{1-\epsilon} + 1 + \frac{3}{\epsilon} \right) t^{1-\epsilon}.$$

Note that for $\sigma \geq \epsilon = 1/2$, we can fix $A = 9$ and for $\sigma \geq \epsilon = 1/4$, we can fix $A = 15$. $\square$

5.6 The Function $\zeta(s) - \frac{1}{s-1}$ is Holomorphic on $Re(s) > 0$

Lemma 5.5. *The function* $\zeta(s) - \frac{1}{s-1}$ *is holomorphic on* $\{s \in \mathbb{C} : Re(s) > 0\}$.

Proof #1. It will be useful to first note two easily verified identities

$$\int_0^\infty x^{-s} \, dx = \frac{1}{s-1} \qquad \text{for } Re(s) > 1, \text{ and}$$

$$\int_n^x \frac{s}{u^{s+1}} \, du = \frac{1}{n^s} - \frac{1}{x^s} \qquad \text{for } Re(s) > 0, n \in \mathbb{N}.$$

Thus, for $Re(s) > 1$, we have

$$\zeta(s) - \frac{1}{s-1} = \sum_{n=1}^\infty \frac{1}{n^s} - \int_0^\infty x^{-s} \, dx = \sum_{n=1}^\infty \int_n^{n+1} \left(\frac{1}{n^s} - \frac{1}{x^s} \right) dx.$$

We stop to define

$$\phi_n(s) = \int_n^{n+1} \left(\frac{1}{n^s} - \frac{1}{x^s} \right) dx \quad \text{and} \quad \phi(s) = \sum_{n=1}^\infty \phi_n(s) \quad \text{so that} \quad \zeta(s) - \frac{1}{s-1} = \phi(s).$$

We want to show that $\phi(s)$ converges absolutely for $Re(s) > 0$. To do so, we first study $\phi_n(s)$, where we know $Re(s) > 0$ and $x \in [n, n+1)$. Applying the ML inequality, we have

$$|\phi_n(s)| \leq 1 \cdot |n^{-s} - x^{-s}| = \left| \int_n^x \frac{s}{u^{s+1}} \, du \right| \leq \int_n^x \frac{|s|}{u^{1+Re(s)}} \, du \leq [(n+1) - n] \frac{|s|}{n^{1+Re(s)}} = \frac{|s|}{n^{1+Re(s)}}.$$

Clearly, each $\phi_n(s)$ is holomorphic for $Re(s) > 0$ and $n \in \mathbb{N}$. That allows

$$|\phi(s)| \leq \sum_{n=1}^{\infty} \frac{|s|}{n^{1+Re(s)}} = |s| \sum_{n=1}^{\infty} \frac{1}{n^{1+Re(s)}}$$

showing that $\phi(s)$ is holomorphic on $Re(s) > 0$ and therefore so is $\zeta(s) - \frac{1}{s-1}$. □

Proof #2. We begin with the results of lemma 5.4, valid for $Re(s) > 0$:

$$\zeta(s) = \frac{s}{s-1} - s \int_1^{\infty} \{x\} x^{-s-1} \, dx = \frac{s}{s-1} - s \int_1^{\infty} \frac{x - [x]}{x^{s+1}} \, dx$$

$$= \frac{s}{s-1} + s \int_1^{\infty} \frac{[x] - x + \frac{1}{2}}{x^{s+1}} \, dx - \int_1^{\infty} \frac{\frac{1}{2}}{x^{s+1}} \, dx$$

$$= \frac{s}{s-1} - \frac{s-1}{s-1} + 1 + s \int_1^{\infty} \frac{[x] - x + \frac{1}{2}}{x^{s+1}} \, dx - \frac{s}{2} \int_1^{\infty} \frac{1}{x^{s+1}} \, dx$$

$$= \frac{1}{s-1} + 1 + s \int_1^{\infty} \frac{[x] - x + \frac{1}{2}}{x^{s+1}} \, dx - \frac{s}{2} \left[\frac{1}{sx^s} \Big|_1^{\infty} \right]$$

$$= \frac{1}{s-1} + s \int_1^{\infty} \frac{[x] - x + \frac{1}{2}}{x^{s+1}} \, dx + \frac{1}{2}$$

$$\zeta(s) - \frac{1}{s-1} = s \int_1^{\infty} \frac{[x] - x + \frac{1}{2}}{x^{s+1}} \, dx + \frac{1}{2}.$$

Since $[x] - x + \frac{1}{2}$ is bounded, the integral converges uniformly and is holomorphic for $Re(s) > 0$, and therefore so is $\zeta(s) - \frac{1}{s-1}$. In particular, we can see that the pole at $s = 1$ is removed by considering[16]

$$\lim_{s \to 1} \left[\zeta(s) - \frac{1}{s-1} \right] = \int_1^{\infty} \frac{[x] - x + \frac{1}{2}}{x^2} \, dx + \frac{1}{2}$$

$$= \lim_{n \to \infty} \left[\int_1^n \frac{[x] - x}{x^2} \, dx + \frac{1}{2} \int_1^n \frac{1}{x^2} \, dx \right] + \frac{1}{2}$$

$$= \lim_{n \to \infty} \int_1^n \frac{[x] - x}{x^2} \, dx + 1 = \lim_{n \to \infty} \left[\int_1^n \frac{[x]}{x^2} \, dx - \log n \right] + 1$$

$$= \lim_{n \to \infty} \left[\sum_{m=1}^{n-1} m \int_m^{m+1} \frac{dx}{x^2} \, dx - \log n \right] + 1$$

$$= \lim_{n \to \infty} \left[\sum_{m=1}^{n-1} \frac{1}{m+1} + 1 - \log n \right] = \lim_{n \to \infty} \left[\sum_{m=1}^{n} \frac{1}{m} - \log n \right] = \gamma,$$

where γ is the Euler-Mascheroni constant. □

Chapter 6

Functional Equation (Contour Method)

6.1 Introduction

In this chapter, we provide a proof of the functional equation for the zeta function. This proof is a detailed version of the *contour method* proof sketched in Riemann's Paper.[17]

6.2 Define Three Contours of Integration

We define three contours of integration that will be used in this proof.

Let $n \in \mathbb{N}$. For small positive δ, we first define our reference point p_2 and then define eight points (including p_2) in the complex plane.

$$p_2 = \delta e^{i\delta} = \delta_x + i\delta_y \quad \text{where} \quad \delta_x = \delta \cos(\delta), \ \delta_y = \delta \sin(\delta)$$

Our first four points are just off the real axis:

$$p_1(n) = \pi(2n+1) + i\delta_y \qquad\qquad p_2 = \delta e^{i\delta} = \delta_x + i\delta_y$$
$$p_3 = \delta e^{-i\delta} = \delta_x - i\delta_y \qquad\qquad p_4(n) = \pi(2n+1) - i\delta_y$$

Our other four points are the four corners of a square centered at the origin:

$$p_5(n) = \pi(2n+1) + i\pi(2n+1) \qquad\qquad p_6(n) = -\pi(2n+1) + i\pi(2n+1)$$
$$p_7(n) = -\pi(2n+1) - i\pi[(2n+1)] \qquad\qquad p_8(n) = \pi(2n+1) - i\pi(2n+1)$$

Using those points, we define the following paths:

- $P_{RE1}(n)$ follows a path parallel to (and just above) the real axis from $+\infty$ to $p_1(n)$.
- $P_{RE4}(n)$ follows a path parallel to (and just below) the real axis from $p_4(n)$ to $+\infty$.
- $P_{CIR}(n)$ follows a path: (1) a straight line from $p_1(n)$ to p_2, (2) then a counterclockwise circular path from p_2 to p_3, tracing the boundary of a circle centered at 0 with radius of δ (making just barely less than a full circle), and (3) then a straight line from p_3 to $p_4(n)$.
- $P_{SQ}(n)$ follows a path $[p_1(n) \to p_5(n) \to p_6(n) \to p_7(n) \to p_8(n) \to p_4(n)]$, of five connected straight lines tracing the boundary of a square (making just barely less than a full square).

Using the above paths, we define two contours:

- $C_H(n)$ follows the counterclockwise path $P_{RE1}(n) + P_{CIR}(n) + P_{RE4}(n)$.
- $C_{SQ}(n)$ follows the counterclockwise path $P_{RE1}(n) + P_{SQ}(n) + P_{RE4}(n)$.

Now we define the closed contour $C_{CL}(n)$ that is created by subtracting $C_H(n)$ from $C_{SQ}(n)$:

$$\begin{aligned}
C_{CL}(n) &= C_{SQ}(n) - C_H(n) \\
&= P_{RE1}(n) + P_{SQ}(n) + P_{RE4}(n) - \left[P_{RE1}(n) + P_{CIR}(n) + P_{RE4}(n)\right] \\
&= P_{SQ}(n) - P_{CIR}(n)
\end{aligned}$$

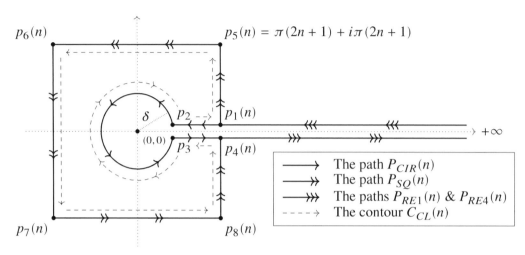

Figure 6.1: Paths and Contour of Integration

We see that $C_{CL}(n)$ is the closed contour created by following $P_{SQ}(n)$ (from $p_1(n)$ to $p_4(n)$) and then traversing the path $P_{CIR}(n)$, but in the clockwise direction (from $p_4(n)$ to $p_1(n)$).

Note that $C_H(n)$ is the same contour as $C_H(n+1)$. The only difference between the two is the actual location of the "junction points" that connect the path $P_{CIR}(n)$ with the paths $P_{RE1}(n)$ and $P_{RE4}(n)$. Thus:

$$C_H(n) = C_H(n+1) = \lim_{n\to\infty} C_H(n).$$

Below, we will simply refer to this contour as C_H. Finally, as a key takeaway from these definitions:

$$\lim_{n\to\infty} C_{CL}(n) = \lim_{n\to\infty} C_{SQ}(n) - \lim_{n\to\infty} C_H(n) = \lim_{n\to\infty} C_{SQ}(n) - C_H. \tag{6.1}$$

6.3 Analyze $|e^z - 1|$ on the Path $P_{SQ}(n)$

Lemma 6.1. *Let $z = x + iy \in \mathbb{C}$ and $n \in \mathbb{N}$. For all points on $P_{SQ}(n)$, $|e^z - 1| > \frac{1}{2}$.*

Proof. We consider the two vertical edges and the two horizontal edges of the square:
The right vertical edge. We have $|e^z - 1| \geq |e^z| - |1| = (e^x - 1) \geq \left(e^{3\pi} - 1\right) > 1$.
The left vertical edge. We have $|e^z - 1| = |1 - e^z| \geq |1| - |e^z| = (1 - e^x) \geq \left(1 - e^{-3\pi}\right) > \frac{1}{2}$.
The top and bottom horizontal edges. We have $y = \pm\pi(2n+1)$. But e^z is periodic, so e^z is unchanged when you substitute $y = \pi$ for $y = \pm\pi(2n+1)$. So, $e^z = e^{x+iy} = e^x e^{iy} = e^x e^{i\pi} = -e^x$. Thus, e^z is real-valued and negative for all z on either horizontal edge, and therefore $|e^z - 1| \geq 1$. $\qquad\square$

6.4 Analyze an Integral over the Open Contour $C_{SQ}(n)$

Lemma 6.2. *Fix* $\{s \in \mathbb{C} : Re(s) < 0\}$. *Then:*

$$\lim_{n \to \infty} \int_{C_{SQ}(n)} \frac{-z^{s-1}}{e^z - 1} \, dz = 0.$$

Proof. From Lemma 6.1 we have on the edges of the square $|e^z - 1| > \frac{1}{2}$. Therefore:

$$\left| \frac{-z^{s-1}}{e^z - 1} \right| < \left| \frac{-z^{s-1}}{1/2} \right| = 2 \left| -z^{s-1} \right|.$$

That means the integrand is bound on the edges of the square by $[2\pi(2n+1)]^{s-1}$. The length of the three longest sides of the square equals $2\pi(2n+1)$. Thus, by the M L estimate, the combined integral over the four sides of the square is bounded by $4 [2\pi(2n+1)]^s$. Since $Re(s) < 0$, this tends to 0 as $n \to \infty$.

That leaves only the "tail pieces" defined by paths $P_{RE1}(n)$ and $P_{RE4}(n)$. With $Re(s) < 0$, we study the numerator and denominator to see that the integrand quickly approach 0 as $n \to \infty$. Sketching an actual proof, we can let n stand in for $2\pi(2n+1)$ (which merely slows the race to infinity).

For the denominator, $\left| e^{n \pm i \delta_y} - 1 \right| \geq \left| e^{n \pm i \delta_y} \right| - |1| = e^n - 1 \geq e^{n-1} = e^n/e$.

For the numerator, fix s, let $A = (Re(s) - 1)$, let $B = Im(s)$ and let $R = \sqrt{n^2 + \delta_y^2}$. Using the principal values of the logarithm (argument θ: $-\pi < \theta \leq \pi$), we write $-z = R e^{i\theta} = e^{\log(R) + i\theta}$, so that

$$\left| -z^{s-1} \right| = \left| e^{(\log(R) + i\theta)[A + iB]} \right| = \left| e^{A \log(R) - B\theta} e^{i(B \log(R) + A\theta)} \right| = \left| e^{A \log(R) - B\theta} \right| = R^A e^{-B\theta} \leq R^A e^{\pi|B|}.$$

But $e^{\pi|B|}$ is fixed. Because $A < -1$, we have $R^A \leq 1/R \leq 1/n$ so that $\lim_{n \to \infty} R^A = 0$. Letting $K = e^{\pi|B|} \cdot e$, we see the absolute value of our integrand is $\leq \dfrac{K}{n \cdot e^n}$ for any given n. $\qquad \square$

6.5 Analyze an Integral over the Closed Contour $C_{CL}(n)$

Lemma 6.3. *Fix* $\{s \in \mathbb{C} : Re(s) < 0\}$. *Then:*

$$\lim_{n \to \infty} \int_{C_{CL}(n)} \frac{-z^{s-1}}{e^z - 1} \, dz = (2\pi i) \left[2^s \pi^{s-1} \sin\left(\frac{\pi s}{2} \right) \zeta(1 - s) \right].$$

Proof. In the integral, $C_{CL}(n)$ is a simple closed contour. The only poles of the integrand are at the points z where $Re(z) = 0$ and $Im(z)$ is a multiple of $\pm 2\pi i m$, for $m \in \mathbb{N}$ and $m \leq n$. (Note that we defined our contour to avoid any singularities on its boundary.) Now let $K = \pm 2\pi i m$ and calculate the residue at K:

$$Res(K) = \lim_{z \to K} (z - K) \frac{-z^{s-1}}{e^z - 1} = \lim_{z \to K} \frac{(z - K)(-K^{s-1})}{e^z - 1} = -K^{s-1} \lim_{z \to K} \frac{z - K}{e^z - 1}$$

now by l'Hospital (since $e^K = 1$):

$$= -K^{s-1} \lim_{z \to K} \frac{(z - K)'}{(e^z - 1)'} = -K^{s-1} \lim_{z \to K} \left(\frac{1}{e^z} \right) = -K^{s-1} \left(\frac{1}{e^K} \right) = -K^{s-1}.$$

So, the residue at $2\pi im$ is $(-2\pi im)^{s-1}$ and the residue at $-2\pi im$ is $(2\pi im)^{s-1}$. Adding up all of the residues for $C_{CL}(n)$, we have:

$$
\begin{aligned}
Res(C_{CL}(n)) &= \sum_{m=1}^{n} \left[(-2\pi im)^{s-1} + (2\pi im)^{s-1} \right] = \sum_{m=1}^{n} (2\pi m)^{s-1} \left((-i)^{s-1} + i^{s-1} \right) \\
&= \sum_{m=1}^{n} (2\pi m)^{s-1} \left(i^{s-1} + (-i)^{s-1} \right) = 2 \sum_{m=1}^{n} (2\pi m)^{s-1} \left[(i^s - (-i)^s)/2i \right] \\
&= 2 \sum_{m=1}^{n} (2\pi m)^{s-1} \left[\frac{e^{i\pi s/2} - e^{-i\pi s/2}}{2i} \right] \quad \left(\text{because } i = e^{\frac{\pi i}{2}} \right) \\
&= 2 \sum_{m=1}^{n} (2\pi m)^{s-1} \left[\sin\left(\frac{\pi s}{2} \right) \right] \quad \left(\text{because } \sin(s) = \frac{e^{is} - e^{-is}}{2i} \right) \\
&= 2^s \pi^{s-1} \sin\left(\frac{\pi s}{2} \right) \sum_{m=1}^{n} m^{s-1} = 2^s \pi^{s-1} \sin\left(\frac{\pi s}{2} \right) \sum_{m=1}^{n} \frac{1}{m^{(1-s)}}.
\end{aligned}
$$

The residues for $C_{CL}(n)$ as $n \to \infty$ equal:

$$
\lim_{n \to \infty} Res(C_{CL}(n)) = \lim_{n \to \infty} 2^s \pi^{s-1} \sin\left(\frac{\pi s}{2} \right) \sum_{m=1}^{n} \frac{1}{m^{(1-s)}} = \left[2^s \pi^{s-1} \sin\left(\frac{\pi s}{2} \right) \zeta(1-s) \right].
$$

Applying *Cauchy's residue theorem*, we have our result:

$$
\lim_{n \to \infty} \int_{C_{CL}(n)} \frac{-z^{s-1}}{e^z - 1} \, dz = (2\pi i) \left[2^s \pi^{s-1} \sin\left(\frac{\pi s}{2} \right) \zeta(1-s) \right]. \qquad \square
$$

6.6 Putting it All Together

Theorem 6.1 (Functional Equation – Contour Method). *We have by Theorem 5.1 that the function $\zeta(s)$ is meromorphic on $\mathbb{C}$, with one simple pole at $s = 1$ with residue 1. We show here that $\zeta(s)$ satisfies the functional equation:*

$$
\zeta(s) = \Gamma(1-s) \left[2^s \pi^{s-1} \sin\left(\frac{\pi s}{2} \right) \zeta(1-s) \right].
$$

Proof. We have the following result obtained in Theorem 5.1:

$$
\zeta(s) = \frac{\Gamma(1-s)}{-2\pi i} \int_C \frac{-z^{s-1}}{e^z - 1} \, dz. \tag{6.2}
$$

Let $\{s \in \mathbb{C} : Re(s) < 0\}$. Note that the contour C in the above integral (described beginning at page 39) is identical to C_H as defined in Section 6.2.

From equation (6.1), we know:

$$
\lim_{n \to \infty} \int_{C_{CL}(n)} \frac{-z^{s-1}}{e^z - 1} \, dz = \lim_{n \to \infty} \int_{C_{SQ}(n)} \frac{-z^{s-1}}{e^z - 1} \, dz - \lim_{n \to \infty} \int_{C_H} \frac{-z^{s-1}}{e^z - 1} \, dz.
$$

Applying Lemmas 6.2 and 6.3, we have:

$$(2\pi i) \left[2^s \pi^{s-1} \sin\left(\frac{\pi s}{2}\right) \zeta(1-s) \right] = [0] - \lim_{n\to\infty} \int_{C_H} \frac{-z^{s-1}}{e^z - 1} \, dz,$$

and therefore:

$$\lim_{n\to\infty} \int_{C_H} \frac{-z^{s-1}}{e^z - 1} \, dz = \int_C \frac{-z^{s-1}}{e^z - 1} \, dz = (-2\pi i) \left[2^s \pi^{s-1} \sin\left(\frac{\pi s}{2}\right) \zeta(1-s) \right].$$

Now applying equation (6.2), we have:

$$\zeta(s) = \frac{\Gamma(1-s)}{-2\pi i} \int_C \frac{-z^{s-1}}{e^z - 1} \, dz = \frac{\Gamma(1-s)}{-2\pi i} (-2\pi i) \left[2^s \pi^{s-1} \sin\left(\frac{\pi s}{2}\right) \zeta(1-s) \right]$$
$$= \Gamma(1-s) \left[2^s \pi^{s-1} \sin\left(\frac{\pi s}{2}\right) \zeta(1-s) \right].$$

We have obtained our functional equation, at least for $Re(s) < 0$. But two meromorphic functions which agree on a nonempty open set are identical (the uniqueness principle). Since both sides of our equation are meromorphic on $\mathbb{C}$, the identity holds for all $s \in \mathbb{C}$. We discuss in detail the zeros and poles of $\zeta(s)$ in section 7.6. □

Chapter 7

Functional Equation (Theta Method)

7.1 Introduction

In this chapter, we provide a proof of the functional equation for the zeta function. This proof is a detailed version of the *theta method* proof sketched in Riemann's Paper.

To begin, we consider one of the simplest among the group of Jacobi theta functions:

$$\vartheta(t) = \sum_{n=-\infty}^{\infty} e^{-\pi n^2 t}.$$

We will show that $\vartheta(t)$ is its own Fourier transform. Using that fact, we provide a functional equation for $\vartheta(t)$. Riemann, of course, was aware of the properties of $\vartheta(t)$ and used them at a critical point in this second proof of the functional equation for the zeta function.

Because our proof relies on Fourier analysis (a huge topic), we include a *very brief* discussion of Fourier analysis, Fourier transforms and the Poisson Summation Formula.

7.2 Fourier Analysis - A Glancing Blow

To keep this simple, we will assume the function $f(z)$ is holomorphic for all $z \in \mathbb{C}$. We will further assume that for each $a > 0$ there exists a constant $A > 0$ such that

$$|f(x + iy)| \leq \frac{A}{1 + x^2} \quad \text{for all } x \in \mathbb{R} \text{ and } |y| < a.$$

The last assumption ensures that $f(z)$ is of sufficient decay on each horizontal line $Im(z) = y$.

With the above assumptions, the Fourier transform for $f(z)$ is:

$$\hat{f}(\xi) = \int_{-\infty}^{\infty} f(x)e^{-2\pi i x\xi}\, dx, \quad x \in \mathbb{R},$$

and the related Fourier inversion formula is:

$$f(x) = \int_{-\infty}^{\infty} \hat{f}(\xi)e^{2\pi i x\xi}\, d\xi, \quad x \in \mathbb{R}.$$

Finally, the key takeaway is the Poisson Summation Formula:

$$\sum_{n\in\mathbb{Z}} f(n) = \sum_{n\in\mathbb{Z}} \hat{f}(n).$$

7.3 Our Simple Theta Function

Lemma 7.1. *For $x \in \mathbb{R}$:* $\int_{-\infty}^{\infty} e^{-\pi x^2} \, dx = 1.$

Proof. We use the multiplicative property of the exponential and a two-dimensional integral:

$$\left(\int_{-\infty}^{\infty} e^{-\pi x^2} \, dx \right)^2 = \left(\int_{-\infty}^{\infty} e^{-\pi x^2} \, dx \right) \left(\int_{-\infty}^{\infty} e^{-\pi y^2} \, dy \right) = \int_{-\infty}^{\infty} \int_{-\infty}^{\infty} e^{-\pi (x^2 + y^2)} \, dx \, dy.$$

Now change to polar coordinates (r, θ) where $dx \, dy = r \, dr \, d\theta$ and $(x^2 + y^2) = r^2$:

$$= \int_0^{2\pi} \int_0^{\infty} e^{-\pi r^2} r \, dr \, d\theta = \left(\int_0^{2\pi} d\theta \right) \left(\int_0^{\infty} e^{-\pi r^2} r \, dr \right) = 2\pi \int_0^{\infty} e^{-\pi r^2} r \, dr.$$

Finally, change variables so that $u = r^2$ and $du = 2r \, dr$:

$$= 2\pi \int_0^{\infty} e^{-\pi u} \frac{du}{2} = \pi \int_0^{\infty} e^{-\pi u} \, du = \pi \left[\frac{1}{-\pi} e^{-\pi u} \right]_0^{\infty} = 0 - (-1) = 1. \qquad \square$$

Lemma 7.2. *If $f(x) = e^{-\pi x^2}$, then $\hat{f}(\xi) = f(\xi)$. Stated differently, we must prove:*

$$f(\xi) = e^{-\pi \xi^2} = \hat{f}(\xi) = \int_{-\infty}^{\infty} e^{-\pi x^2} e^{-2\pi i x \xi} \, dx.$$

Proof.[18]
First, assume $\xi = 0$. We use Lemma 7.1 and have:

$$\hat{f}(0) = \int_{-\infty}^{\infty} e^{-\pi x^2} e^{(-2\pi i x) 0} \, dx = \int_{-\infty}^{\infty} e^{-\pi x^2} \, dx = 1.$$

Next, assume $\xi > 0$ and fix ξ.

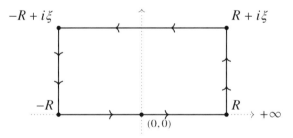

Figure 7.1: The Contour C_R

We define the closed contour C_R, which consists of a rectangle with corners $R, R + i\xi, -R + i\xi, -R$, oriented in the counterclockwise direction. The function $f(z) = e^{-\pi z^2}$ is holomorphic on and in the interior of C_R. Therefore, by Cauchy's integral theorem:

$$\int_{C_R} f(z) \, dz = 0.$$

Now consider the integrals over each of the four sides of the rectangle. The integral over the bottom (horizontal real segment) is:

$$I(R \to) = \int_{-R}^{R} e^{-\pi x^2} dx \qquad \text{so that} \qquad \lim_{R \to \infty} I(R \to) \to 1,$$

by application of lemma 7.1.

The right vertical side consists of points of the form $(R + iy)$ for $0 \le y \le \xi$, so the integral over the right vertical side is:

$$I(R \uparrow) = \int_0^{\xi} f(R + iy) i\, dy = \int_0^{\xi} e^{-\pi(R^2 + 2iRy - y^2)} i\, dy.$$

By the ML inequality, we have:

$$|I(R \uparrow)| \le |\xi| \max_{0 \le y \le \xi} \left| e^{-\pi(R^2 + 2iRy - y^2)} i \right| = |\xi| \max_{0 \le y \le \xi} \left| e^{-\pi(R^2 - y^2)} \right| \le |\xi| e^{\pi|\xi|^2} e^{-\pi R^2}.$$

For our fixed ξ, define $A_\xi = |\xi| e^{\pi|\xi|^2}$. Therefore, A_ξ is also fixed and we have:

$$\lim_{R \to \infty} |I(R \uparrow)| \le \lim_{R \to \infty} A_\xi e^{-\pi R^2} = 0.$$

Using the same logic, the integral over the left side, $I(R \downarrow)$, also goes to 0. For the final integral over the top (horizontal imaginary segment):

$$I(R \leftarrow) = \int_R^{-R} e^{-\pi(x + i\xi)^2} dx = \int_R^{-R} e^{-\pi(x^2 + 2xi\xi - \xi^2)} dx = -e^{\pi\xi^2} \int_{-R}^{R} e^{-\pi x^2} e^{-2\pi i x\xi} dx.$$

So, taken to the limit as $R \to \infty$, we have:

$$\int_{C_R} f(z)\, dz = 0 = [I(R \to) + I(R \uparrow) + I(R \downarrow) + I(R \leftarrow)] = 1 - e^{\pi\xi^2} \int_{-\infty}^{\infty} e^{-\pi x^2} e^{-2\pi i x\xi} dx.$$

which is to say

$$0 = 1 - e^{\pi\xi^2} \int_{-\infty}^{\infty} e^{-\pi x^2} e^{-2\pi i x\xi} dx$$

$$1 = e^{\pi\xi^2} \int_{-\infty}^{\infty} e^{-\pi x^2} e^{-2\pi i x\xi} dx$$

$$e^{-\pi\xi^2} = \int_{-\infty}^{\infty} e^{-\pi x^2} e^{-2\pi i x\xi} dx,$$

as required.

Finally, assume $\xi < 0$ and fix ξ.

We use a symmetric rectangle in the lower half plane to define our new closed contour C_R'. The contour consists of a rectangle with corners $R, R - i\xi, -R - i\xi, -R$, oriented in the *clockwise* direction. We again consider the four sides of the rectangle. $I(R \to)$ along the horizontal real segment is identical to the $\xi > 0$ case so converges to 1 as $R \to \infty$. The analysis for the two vertical sides is also the same (and explains why we used $|\xi|$ above). Finally, the orientation of the remaining horizontal segment (from $R - i\xi$ to $-R - i\xi$) is the same as in the $\xi > 0$ case, so again yields the same result. Our proof is therefore complete. $\qquad\square$

Lemma 7.3. *For $t > 0$ and $\vartheta(t) = \sum_{n=-\infty}^{\infty} e^{-\pi n^2 t}$, we have $\vartheta(t) = \dfrac{1}{\sqrt{t}} \vartheta\left(\dfrac{1}{t}\right)$.*

Proof. From Lemma 7.2, if $f(x) = e^{-\pi x^2}$, then $\hat{f}(\xi) = f(\xi)$, which means:

$$f(\xi) = e^{-\pi \xi^2} = \hat{f}(\xi) = \int_{-\infty}^{\infty} e^{-\pi x^2} e^{-2\pi i x \xi}\, dx.$$

We also note the Poisson Summation Formula:

$$\sum_{n=-\infty}^{\infty} f(n) = \sum_{n=-\infty}^{\infty} \hat{f}(n).$$

Combining the above, we have:

$$\sum_{n=-\infty}^{\infty} e^{-\pi(ns)^2} = \sum_{n=-\infty}^{\infty} \int_{-\infty}^{\infty} e^{-\pi x^2} e^{-2\pi i x(ns)}\, dx = \sum_{n=-\infty}^{\infty} \frac{1}{s} \int_{-\infty}^{\infty} e^{-\pi(u/s)^2} e^{-2\pi i x u}\, du.$$

The last equation above is from a change variables so that $u = sx$ and $du = s\,dx$. Now we can again use Lemma 7.2 to convert the last equation to:

$$= \frac{1}{s} \sum_{n=-\infty}^{\infty} e^{-\pi(n/s)^2}.$$

We have shown:

$$\sum_{n=-\infty}^{\infty} e^{-\pi(ns)^2} = \frac{1}{s} \sum_{n=-\infty}^{\infty} e^{-\pi(n/s)^2}$$

To complete the proof, take $s = \sqrt{t}$:

$$\sum_{n=-\infty}^{\infty} e^{-\pi n^2 t} = \frac{1}{\sqrt{t}} \sum_{n=-\infty}^{\infty} e^{-\pi n^2 / t}.$$ $\square$

7.4 The Problematic Integral

Lemma 7.4. *For $x > 0$, define $w(x) = \sum_{n=1}^{\infty} e^{-\pi n^2 x}$. Then the following integral:*

$$\int_0^1 x^{\left(\frac{s}{2}-1\right)} w(x)\, dx \quad \text{(We will call this the "problematic integral")}$$

is equal to

$$\int_1^{\infty} x^{-\left(\frac{s}{2}+\frac{1}{2}\right)} w(x)\, dx + \frac{1}{s(s-1)}.$$

Proof. Using Lemma 7.3, we have for $x > 0$:

$$\vartheta(x) = \sum_{n=-\infty}^{\infty} e^{-\pi n^2 x} = 1 + 2\sum_{n=1}^{\infty} e^{-\pi n^2 x} = 1 + 2w(x) = \frac{1}{\sqrt{x}} \vartheta\left(\frac{1}{x}\right) = \frac{1}{\sqrt{x}}\left[1 + 2w\left(\frac{1}{x}\right)\right].$$

That allows us to express the $w(x)$ function as follows:

$$\vartheta(x) = \frac{1}{\sqrt{x}}\vartheta\left(\frac{1}{x}\right)$$

$$2w(x) + 1 = \frac{1}{\sqrt{x}}\left[2w\left(\frac{1}{x}\right) + 1\right]$$

$$2w(x) = \frac{1}{\sqrt{x}}\left[2w\left(\frac{1}{x}\right) + 1\right] - 1$$

$$w(x) = \frac{1}{\sqrt{x}}\left[\frac{2}{2}w\left(\frac{1}{x}\right) + \frac{1}{2}\right] - \frac{1}{2} = \frac{1}{\sqrt{x}}w\left(\frac{1}{x}\right) + \frac{1}{2\sqrt{x}} - \frac{1}{2}.$$

Now returning to the problematic integral, we have:

$$\int_0^1 x^{\left(\frac{s}{2}-1\right)}w(x)\,dx = \int_0^1 x^{\left(\frac{s}{2}-1\right)}\left[\frac{1}{\sqrt{x}}w\left(\frac{1}{x}\right) + \frac{1}{2\sqrt{x}} - \frac{1}{2}\right]dx$$

$$= \int_0^1 x^{\left(\frac{s}{2}-1\right)}\left[\frac{1}{\sqrt{x}}w\left(\frac{1}{x}\right)\right]dx + \int_0^1 x^{\left(\frac{s}{2}-1\right)}\left[\frac{1}{2\sqrt{x}} - \frac{1}{2}\right]dx$$

$$= \int_0^1 x^{\left(\frac{s}{2}-\frac{3}{2}\right)}w\left(\frac{1}{x}\right)dx + \frac{1}{2}\int_0^1 x^{\left(\frac{s}{2}-\frac{3}{2}\right)} - x^{\left(\frac{s}{2}-1\right)}\,dx.$$

The right integral can easily be evaluated:

$$\frac{1}{2}\int_0^1 x^{\left(\frac{s}{2}-\frac{3}{2}\right)} - x^{\left(\frac{s}{2}-1\right)}\,dx = \frac{1}{2}\left[\left(\frac{s}{2}-\frac{1}{2}\right)^{-1}x^{\left(\frac{s}{2}-\frac{1}{2}\right)} - \left(\frac{s}{2}\right)^{-1}x^{\left(\frac{s}{2}\right)}\right]_0^1 = \frac{1}{s(s-1)}.$$

So the problematic integral becomes:

$$\int_0^1 x^{\left(\frac{s}{2}-\frac{3}{2}\right)}w\left(\frac{1}{x}\right)dx + \frac{1}{s(s-1)}.$$

Now let

$$x = \frac{1}{u} \text{ and } dx = -\frac{1}{u^2}\,du; \text{ which means } \int_0^1 dx \to \int_\infty^1 du.$$

We have:

$$\int_0^1 x^{\left(\frac{s}{2}-\frac{3}{2}\right)}w\left(\frac{1}{x}\right)dx + \frac{1}{s(s-1)} = \int_\infty^1 \frac{1}{u}^{\left(\frac{s}{2}-\frac{3}{2}\right)}w(u)\frac{-1}{u^2}\,du + \frac{1}{s(s-1)}$$

$$= -\int_\infty^1 \left(\frac{1}{u}\right)^2 \frac{1}{u}^{\left(\frac{s}{2}-\frac{3}{2}\right)}w(u)\,du + \frac{1}{s(s-1)}.$$

Now reversing the direction of integration

$$= \int_1^\infty \frac{1}{u}^{\left(\frac{s}{2}+\frac{1}{2}\right)}w(u)\,du + \frac{1}{s(s-1)}$$

$$= \int_1^\infty u^{-\left(\frac{s}{2}+\frac{1}{2}\right)}w(u)\,du + \frac{1}{s(s-1)}. \qquad \square$$

7.5 Functional Equation of Zeta Function

Theorem 7.1 (Functional Equation – Theta Method). *We have by Theorem 5.1 that the function $\zeta(s)$ is meromorphic on $\mathbb{C}$, with one simple pole at $s = 1$ with residue 1. We show here that $\zeta(s)$ satisfies the functional equation:*

$$\pi^{\frac{-s}{2}} \Gamma\left(\frac{s}{2}\right) \zeta(s) = \pi^{\frac{-(1-s)}{2}} \Gamma\left(\frac{1-s}{2}\right) \zeta(1-s).$$

Proof. We start with the Gamma function:

$$\Gamma(s) = \int_0^\infty e^{-t} t^{(s-1)} \, dt.$$

First, replace s by $s/2$:

$$\Gamma\left(\frac{s}{2}\right) = \int_0^\infty e^{-t} t^{\left(\frac{s}{2}-1\right)} \, dt.$$

Next, let $t = \pi n^2 x$ so that $dt = \pi n^2 \, dx$:

$$\Gamma\left(\frac{s}{2}\right) = \int_0^\infty \left(\pi n^2 x\right)^{\left(\frac{s}{2}-1\right)} e^{-\pi n^2 x} \pi n^2 \, dx$$

$$= \int_0^\infty \pi^{\frac{s}{2}} \left(n^2\right)^{\frac{s}{2}} x^{\left(\frac{s}{2}-1\right)} e^{-\pi n^2 x} \, dx$$

$$= \int_0^\infty \pi^{\frac{s}{2}} n^s x^{\left(\frac{s}{2}-1\right)} e^{-\pi n^2 x} \, dx.$$

With some rearrangement, we have:

$$\pi^{\frac{-s}{2}} \Gamma\left(\frac{s}{2}\right) \frac{1}{n^s} = \int_0^\infty x^{\left(\frac{s}{2}-1\right)} e^{-\pi n^2 x} \, dx.$$

Now summing both sides and rearranging terms:

$$\sum_{n=1}^\infty \left[\pi^{\frac{-s}{2}} \Gamma\left(\frac{s}{2}\right) \frac{1}{n^s} \right] = \sum_{n=1}^\infty \left[\int_0^\infty x^{\left(\frac{s}{2}-1\right)} e^{-\pi n^2 x} \, dx \right]$$

$$\pi^{\frac{-s}{2}} \Gamma\left(\frac{s}{2}\right) \sum_{n=1}^\infty \frac{1}{n^s} = \sum_{n=1}^\infty \left[\int_0^\infty x^{\left(\frac{s}{2}-1\right)} e^{-\pi n^2 x} \, dx \right]$$

$$\pi^{\frac{-s}{2}} \Gamma\left(\frac{s}{2}\right) \zeta(s) = \int_0^\infty x^{\left(\frac{s}{2}-1\right)} \sum_{n=1}^\infty e^{-\pi n^2 x} \, dx.$$

Set $w(x) = \sum_{n=1}^\infty e^{-\pi n^2 x}$, giving:

$$\pi^{\frac{-s}{2}} \Gamma\left(\frac{s}{2}\right) \zeta(s) = \int_0^\infty x^{\left(\frac{s}{2}-1\right)} w(x) \, dx$$

$$= \int_0^1 x^{\left(\frac{s}{2}-1\right)} w(x) \, dx + \int_1^\infty x^{\left(\frac{s}{2}-1\right)} w(x) \, dx.$$

We divided the integral into two parts because the 1 to ∞ integral is well behaved but the 0 to 1 integral is problematic. The trick is to use the reciprocal version of the theta function on the problematic part. We have done that in Lemma 7.4, and so can restate the above equality with the Lemma 7.4 substitution for the problematic integral:

$$\pi^{\frac{-s}{2}}\Gamma\left(\frac{s}{2}\right)\zeta(s) = \int_0^1 x^{\left(\frac{s}{2}-1\right)}w(x)\,dx + \int_1^\infty x^{\left(\frac{s}{2}-1\right)}w(x)\,dx$$

$$= \int_1^\infty x^{-\left(\frac{s}{2}+\frac{1}{2}\right)}w(x)\,dx + \frac{1}{s(s-1)} + \int_1^\infty x^{\left(\frac{s}{2}-1\right)}w(x)\,dx.$$

Simplifying:

$$= \frac{1}{s(s-1)} + \int_1^\infty \left[x^{\left(\frac{-s}{2}-\frac{1}{2}\right)} + x^{\left(\frac{s}{2}-1\right)}\right]w(x)\,dx. \tag{7.1}$$

Note that the right-hand side of the last equation is unchanged when you replace s by $(1-s)$. Therefore, so long as the integral is convergent for all s, both sides are meromorphic on $\mathbb{C}$ and we have our functional equation:

$$\pi^{\frac{-s}{2}}\Gamma\left(\frac{s}{2}\right)\zeta(s) = \pi^{\frac{-(1-s)}{2}}\Gamma\left(\frac{1-s}{2}\right)\zeta(1-s). \tag{7.2}$$

Regarding convergence of the integral, we observe that:

$$w(x) = \sum_{n=1}^\infty e^{-\pi n^2 x} < \sum_{n=1}^\infty e^{-\pi n x} = \frac{1}{e^{\pi x}-1}$$

and note that $e^{\pi x}$ grows more rapidly than any power of x. $\qquad\square$

Corollary 7.1. *(Re-Write Functional Equation): The functional equation can be re-written as:*

$$\zeta(s) = 2^s \pi^{(s-1)}\sin\left(\frac{\pi s}{2}\right)\Gamma(1-s)\zeta(1-s).$$

Proof. We first note two special formulae of the Gamma function:

1. *Euler's Reflection Formula:* $\Gamma(s)\Gamma(1-s) = \dfrac{\pi}{\sin(\pi s)}.$

2. *Legendre Duplication Formula:* $\Gamma(2s) = \Gamma(s)\Gamma(s+\frac{1}{2})\pi^{-\frac{1}{2}}2^{(2s-1)}.$

We make some simple substitutions and rearrangements (the final form will suit our needs):

$$\Gamma(2s) = \Gamma(s)\Gamma\left(s+\frac{1}{2}\right)\pi^{-\frac{1}{2}}2^{(2s-1)}$$

$$\Gamma(s) = \Gamma\left(\frac{s}{2}\right)\Gamma\left(\frac{s+1}{2}\right)\pi^{-\frac{1}{2}}2^{(s-1)}$$

$$\Gamma(1-s) = \Gamma\left(\frac{1-s}{2}\right)\Gamma\left(\frac{2-s}{2}\right)\pi^{-\frac{1}{2}}2^{(-s)}$$

$$\pi^{\frac{1}{2}}2^s\Gamma(1-s)\frac{1}{\Gamma\left(\frac{1-s}{2}\right)} = \Gamma\left(\frac{1-s}{2}\right).$$

Now we start with our previously proved functional equation:

$$\pi^{\frac{-s}{2}} \Gamma\left(\frac{s}{2}\right) \zeta(s) = \pi^{\frac{-(1-s)}{2}} \left[\Gamma\left(\frac{1-s}{2}\right)\right] \zeta(1-s).$$

Our first step is to substitute the final form above of the Legendre Duplication Formula in our equation:

$$\pi^{\frac{-s}{2}} \Gamma\left(\frac{s}{2}\right) \zeta(s) = \pi^{\frac{-(1-s)}{2}} \left[\pi^{\frac{1}{2}} 2^s \Gamma(1-s) \frac{1}{\Gamma\left(\frac{1-s}{2}\right)}\right] \zeta(1-s).$$

Next, some rearrangements:

$$\pi^{\frac{-s}{2}} \Gamma\left(\frac{s}{2}\right) \zeta(s) = \pi^{\frac{1}{2}} 2^s \pi^{\frac{-(1-s)}{2}} \left[\frac{1}{\Gamma\left(\frac{1-s}{2}\right)}\right] \Gamma(1-s) \zeta(1-s)$$

$$\zeta(s) = \pi^{\frac{s}{2}} \pi^{\frac{1}{2}} 2^s \pi^{\frac{-(1-s)}{2}} \left[\frac{1}{\Gamma\left(\frac{s}{2}\right)\Gamma\left(\frac{1-s}{2}\right)}\right] \Gamma(1-s) \zeta(1-s)$$

$$\zeta(s) = 2^s \pi^s \left[\frac{1}{\Gamma\left(\frac{s}{2}\right)\Gamma\left(\frac{1-s}{2}\right)}\right] \Gamma(1-s) \zeta(1-s)$$

We have maneuvered the brackets to contain a reciprocal of *Euler's Reflection Formula*. We make the substitution and thereby obtain our desired result:

$$\zeta(s) = 2^s \pi^s \left[\frac{sin\left(\frac{\pi s}{2}\right)}{\pi}\right] \Gamma(1-s) \zeta(1-s)$$

$$= 2^s \pi^{s-1} \left[sin\left(\frac{\pi s}{2}\right)\right] \Gamma(1-s) \zeta(1-s).$$

$$= 2^s \pi^{s-1} sin\left(\frac{\pi s}{2}\right) \Gamma(1-s) \zeta(1-s). \qquad \square$$

7.6   The Zeros/Poles of the Functional Equation.

7.6.1   The Zeros

Let's review our rearranged functional equation: $\zeta(s) = 2^s \pi^{s-1} sin\left(\frac{\pi s}{2}\right) \Gamma(1-s) \zeta(1-s).$

- $Re(s) > 1$. There are no zeros for the simple reason that $\sum_{n=1}^{\infty} \frac{1}{n^s}$ converges.

- $Re(s) < 0$. All terms on the RHS are always non-zero, except for the sin() function, which is equal to 0 at the negative even integers. We have $[sin(\pi s/2) = 0] \Rightarrow [\zeta(s) = 0]$ for $s \in \{-2, -4, -6, ...\}$. These are known as the *trivial zeros*.

- $0 \le Re(s) \le 1$. The zeros cannot be determined by immediate inspection of the above functional equation. By calculation, we know such zeros exist. These are known as the *nontrivial zeros*.

7.6.2 The Poles

Regarding any possible poles, consider equation (5.3): $\quad \zeta(s) = \dfrac{\Gamma(1-s)}{-2\pi i} \displaystyle\int_C \dfrac{-z^{s-1}}{e^z - 1}\, dz.$

The integral is clearly holomorphic for all $s \in \mathbb{C}$. On the other hand, $\Gamma(1-s)$ is meromorphic, with poles at the positive integers. But $\zeta(s)$ is known to be holomorphic for all $Re(s) > 1$, so the poles of $\Gamma(1-s)$ at integers $n \geq 2$ must cancel against zeros of the integral. That leaves one simple pole at $s = 1$ (simple because $\Gamma(1-s)$ is a simple pole). To compute the residue of $\zeta(s)$ at $s = 1$ we must compute $\lim\limits_{s \to 1}(s-1)\zeta(s)$.

Lemma 7.5. *The residue of $\zeta(s)$ at $s = 1$ is equal to* 1.

Proof #1 – Using Equation (5.3). We start again with equation (5.3), above. For $s = 1$ the contour integral is

$$\int_C \frac{dz}{e^z - 1} = \int_\gamma \frac{dz}{e^z - 1} = 2\pi i,$$

where γ is a contour enclosing only the pole at $z = 0$. So, we have:

$$\lim_{s \to 1}(s-1)\zeta(s) = \lim_{s \to 1}(s-1)\frac{\Gamma(1-s)}{-2\pi i}\int_C \frac{-z^{s-1}}{e^z - 1}\, dz$$

$$= \lim_{s \to 1}(s-1)\frac{\Gamma(1-s)}{-2\pi i}\int_C \frac{dz}{e^z - 1} = \lim_{s \to 1}(s-1)\frac{\Gamma(1-s)}{-2\pi i}2\pi i$$

$$= \lim_{s \to 1}(s-1)(-1)\Gamma(1-s) = \lim_{s \to 1}(1-s)\Gamma(1-s)$$

$$= \lim_{s \to 1}(1-s)\frac{\Gamma(1-s+1)}{(1-s)} \qquad \text{(using the functional equation)}$$

$$= \lim_{s \to 1}\Gamma(2-s) = \Gamma(1) = 1. \qquad\qquad \square$$

Proof #2 – Using Real-Valued Zeta Function. For $\{s \in \mathbb{C} : Re(s) > 1\}$ we have

$$\zeta(s) = \sum_{n=1}^{\infty} \frac{1}{n^s} \quad \text{with} \quad \left|\frac{1}{n^s}\right| = \frac{1}{n^{Re(s)}}.$$

The last equality allows us to assume s is real and approach 1 "from the right" along the real line. Let $\epsilon = Re(s) - 1$ so that $s = (1 + \epsilon)$ and $(s - 1) = \epsilon$. Thus, we can evaluate

$$\lim_{s \to 1}(s-1)\zeta(s) \quad \text{or, equally} \quad \lim_{\epsilon \to 0} \epsilon \cdot \zeta(1+\epsilon).$$

We start with these key inequalities

$$\boxed{\zeta(1+\epsilon)} = \sum_{n=1}^{\infty} \frac{1}{n^{1+\epsilon}} = 1 + \sum_{n=1}^{\infty} \frac{1}{(n+1)^{1+\epsilon}} \overset{\leq}{} 1 + \sum_{n=1}^{\infty} \int_n^{n+1} \frac{1}{x^{1+\epsilon}}\, dx = 1 + \int_1^{\infty} \frac{1}{x^{1+\epsilon}}\, dx = \boxed{1 + \frac{1}{\epsilon}}$$

$$= \sum_{n=1}^{\infty} \frac{1}{n^{1+\epsilon}} = \sum_{n=1}^{\infty} \int_n^{n+1} \frac{1}{n^{1+\epsilon}}\, dx \overset{\geq}{} \sum_{n=1}^{\infty} \int_n^{n+1} \frac{1}{x^{1+\epsilon}}\, dx = \int_1^{\infty} \frac{1}{x^{1+\epsilon}}\, dx \qquad = \boxed{\frac{1}{\epsilon}}$$

Putting the inequalities together, and multiplying by ϵ, we have $1 \leq \epsilon \cdot \zeta(1+\epsilon) \leq \epsilon + 1$. Thus, by the squeeze test, $\lim\limits_{\epsilon \to 0} \epsilon \cdot \zeta(1+\epsilon) = 1.$ $\qquad\qquad \square$

Proof #3 – Using Equation (5.4). We start with equation (5.4):

$$\zeta(s) = \frac{s}{s-1} - s\int_1^\infty \frac{\{x\}}{x^{s+1}}\,dx, \quad \text{valid for } Re(s) > 0.$$

Because $0 \le \{x\} \le 1$, the last integral converges and is holomorphic on $Re(s) > 0$. Now consider the value of that integral for $s = 1$

$$s\int_1^\infty \frac{\{x\}}{x^{s+1}}\,dx = \int_1^\infty \frac{\{x\}}{x^2}\,dx \le \int_1^\infty \frac{1}{x^2}\,dx = \left.\frac{-1}{x}\right|_1^\infty = 1.$$

That allows

$$\lim_{s\to 1}\left[(s-1)\zeta(s) = (s-1)\frac{s}{(s-1)} - (s-1)\left(s\int_1^\infty \frac{\{x\}}{x^{s+1}}\,dx\right)\right] = 1 - 0 = 1. \qquad \square$$

Proof #4 – Using Equation (8.1). We start with equation (8.1): $\xi(s) = \frac{s(s-1)}{2}\pi^{\frac{-s}{2}}\Gamma\left(\frac{s}{2}\right)\zeta(s)$. Rearranging, substituting and taken to the limit

$$\lim_{s\to 1}(s-1)\zeta(s) = \lim_{s\to 1}(s-1)\frac{2}{s(s-1)}\frac{\xi(s)}{\pi^{\frac{-s}{2}}\Gamma\left(\frac{s}{2}\right)} = \frac{2}{1}\cdot\frac{\xi(1)}{\pi^{\frac{-1}{2}}\pi^{\frac{1}{2}}} = 2\cdot\xi(1) = 1. \qquad \square$$

7.6.3   The Value of $\zeta(0)$

We start with the final form of the functional equation:

$$\zeta(s) = 2^s\pi^{s-1}\sin\left(\frac{\pi s}{2}\right)\Gamma(1-s)\zeta(1-s).$$

A special calculation of $\zeta(0)$ is needed, because the right sides of the equation blow up at $\zeta(1-0)$. We multiply both sides by $(1-s)$ and take the limit as $s \to 1$:

$$\lim_{s\to 1}(1-s)\zeta(s) = \lim_{s\to 1}(1-s)2^s\pi^{s-1}\sin\left(\frac{\pi s}{2}\right)\Gamma(1-s)\zeta(1-s).$$

By the residue calculus $\lim_{s\to 1}(s-1)\zeta(s) = 1$, so $\lim_{s\to 1}(1-s)\zeta(s) = -1$. Simplifying the above (and noting that $(1-s)\Gamma(1-s) = \Gamma(2-s)$):

$$
\begin{aligned}
-1 &= \lim_{s\to 1}(1-s)2^1\pi^{1-1}\sin\left(\frac{\pi}{2}\right)\Gamma(1-s)\zeta(1-s)\\
&= \lim_{s\to 1}\left[2^1\pi^{1-1}\sin\left(\frac{\pi}{2}\right)\right](1-s)\Gamma(1-s)\zeta(1-s)\\
&= \lim_{s\to 1}[2\cdot 1\cdot 1](1-s)\Gamma(1-s)\zeta(1-s)\\
&= \lim_{s\to 1}2\cdot\Gamma(2-s)\zeta(1-s) = 2\cdot 1\cdot\zeta(0) = 2\cdot\zeta(0).
\end{aligned}
$$

Thus, $\zeta(0) = -\frac{1}{2}$.

Completed Zeta Function

8.1 The Completed Zeta Function.

Define a new function, $\Lambda(s)$, and set it equal to the left side of our functional equation (7.2):

$$\Lambda(s) = \pi^{\frac{-s}{2}} \Gamma\left(\frac{s}{2}\right) \zeta(s)$$

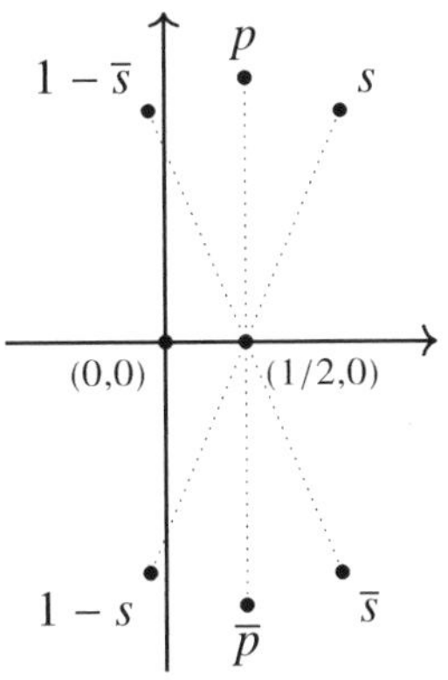

Figure 8.1: Symmetry

Then we have: $\Lambda(s) = \Lambda(1-s)$. This is just another way of stating the functional equation. Some authors (I believe incorrectly) call this the *Completed Zeta Function*. For our *Completed Zeta Function*, we follow Riemann (without his unnecessary change of variables) and define $\xi(s)$ as follows:

$$\xi(s) = \frac{s(s-1)}{2}\Lambda(s) = \frac{s(s-1)}{2}\pi^{\frac{-s}{2}}\Gamma\left(\frac{s}{2}\right)\zeta(s). \qquad (8.1)$$

Because $s(s-1)/2$ is unchanged by the substitution $s \to (1-s)$, we have $\xi(s) = \xi(1-s)$. There are two advantages of $\xi(s)$. First, we have removed the pole at $s = 1$, making $\xi(s)$ an entire function. Second, we have eliminated the *nontrivial zeros* because $\xi(s) = \xi(1-s)$ and $\xi(s) \neq 0$ for $s > 1$. By inspection, the zeros of $\xi(s)$ are precisely the non-trivial zeros of $\zeta(s)$.

With $\xi(s) = \xi(1-s)$, we also have $\xi(\bar{s}) = \xi(1-\bar{s})$. On Figure 8.1, the pairs s and $(1-s)$ and the pairs $\bar{s}$ and $(1-\bar{s})$ are mirror reflections, centered at $(1/2, 0)$. It is easily verified that $\xi(s)$ is real-valued when $Im(s) = 0$. By the Schwarz reflection principle, that means $\xi(\bar{s}) = \overline{\xi(s)}$.

Now consider our point p on the line $Re(s) = 1/2$. A simple calculation shows that $\bar{p} = 1 - p$, so that $\xi(p) = \xi(1-p) = \xi(\bar{p})$. By Schwartz, $\xi(\bar{p}) = \overline{\xi(p)}$, so that $\xi(p) = \overline{\xi(p)}$, which is only possible if $\xi(p)$ is real valued. Thus, $\xi(s) \in \mathbb{R}$ for all $Re(s) = 1/2$.

8.2 The Completed Zeta Function: Another Form.

Riemann developed a wholly different equation for $\xi(s)$. We will follow the proof in Edwards [13, 16-17].

Theorem 8.1. *Let*

$$\xi(s) = \frac{s(s-1)}{2}\pi^{\frac{-s}{2}}\Gamma\left(\frac{s}{2}\right)\zeta(s) \quad and \quad w(x) = \sum_{n=1}^{\infty} e^{-\pi n^2 x}.$$

Further, let

$$a_{2n} = 4 \int_1^\infty \left[\frac{d}{dx} x^{\left(\frac{3}{2}\right)} w'(x) \right] \left(x^{-1/4} \frac{\left(\frac{1}{2}\log(x)\right)^{2n}}{(2n)!} \right) dx.$$

Then we have

$$\xi(s) = \sum_{n=0}^\infty a_{2n} \left(s - \frac{1}{2} \right)^{2n}.$$

Proof. Riemann starts with the right side of equation (7.1), which was used in proving the functional equation:

$$\pi^{\frac{-s}{2}} \Gamma\left(\frac{s}{2}\right) \zeta(s) = \frac{1}{s(s-1)} + \int_1^\infty \left[x^{\left(\frac{-s}{2} - \frac{1}{2}\right)} + x^{\left(\frac{s}{2} - 1\right)} \right] w(x)\, dx,$$

so that

$$\xi(s) = \frac{s(s-1)}{2} \pi^{\frac{-s}{2}} \Gamma\left(\frac{s}{2}\right) \zeta(s) = \frac{s(s-1)}{2} \left[\frac{1}{s(s-1)} + \int_1^\infty \left[x^{\left(\frac{-s}{2} - \frac{1}{2}\right)} + x^{\left(\frac{s}{2} - 1\right)} \right] w(x)\, dx \right]$$

$$= \frac{1}{2} + \frac{s(s-1)}{2} \left[\int_1^\infty \left[x^{\left(\frac{-s}{2} - \frac{1}{2}\right)} + x^{\left(\frac{s}{2} - 1\right)} \right] w(x)\, dx \right]$$

$$= \frac{1}{2} - \frac{s(1-s)}{2} \int_1^\infty \left[x^{\frac{(1-s)}{2} - 1} + x^{\left(\frac{s}{2}\right) - 1} \right] w(x)\, dx.$$

Now let:

$$f(x) = \left[x^{\frac{(1-s)}{2} - 1} + x^{\left(\frac{s}{2}\right) - 1} \right] \text{ and } F(x) = \int f(x) = \frac{x^{\frac{(1-s)}{2}}}{(1-s)/2} + \frac{x^{\left(\frac{s}{2}\right)}}{s/2}.$$

So, we have:

$$\xi(s) = \frac{1}{2} - \frac{s(1-s)}{2} \int_1^\infty f(x) w(x)\, dx.$$

We will use the following version of integration by parts:

$$\int_1^\infty w(x) f(x)\, dx = w(x) \int_1^\infty f(x)\, dx - \int_1^\infty \left[w'(x) \int_1^\infty f(x)\, dx \right] dx$$

$$= \int_1^\infty \frac{d}{dx} \{ w(x) F(x) \}\, dx - \int_1^\infty w'(x) F(x)\, dx.$$

Back to Edwards, we have:

$$\xi(s) = \frac{1}{2} - \frac{s(1-s)}{2} \int_1^\infty w(x) f(x)\, dx$$

$$= \frac{1}{2} - \frac{s(1-s)}{2} \left[\int_1^\infty \frac{d}{dx} \{ w(x) F(x) \}\, dx - \int_1^\infty w'(x) F(x)\, dx \right]$$

$$= \frac{1}{2} - \left[\frac{s(1-s)}{2} \int_1^\infty \frac{d}{dx} \{ w(x) F(x) \}\, dx \right] + \left[\frac{s(1-s)}{2} \int_1^\infty w'(x) F(x)\, dx \right].$$

Next, we evaluate the first bracketed integral. We have:

$$\frac{s(1-s)}{2} \int_1^\infty \frac{d}{dx} \{ w(x) F(x) \}\, dx = \frac{s(1-s)}{2} \cdot \left[w(x) F(x) \Big|_1^\infty \right] = \frac{s(1-s)}{2} \cdot \left[\lim_{x \to \infty} w(x) F(x) - w(1) F(1) \right].$$

CHAPTER 8. COMPLETED ZETA FUNCTION

But

$$w(x) = \sum_{n=1}^{\infty} e^{-\pi n^2 x} < \sum_{n=1}^{\infty} e^{-\pi n x} = \frac{1}{e^{\pi x} - 1}$$

and e^x grows faster than any power of x so: $\lim_{x \to \infty} w(x)F(x) = 0$. Therefore, the first integral equals:

$$= \frac{s(1-s)}{2} \cdot [0 - w(1)F(1)] = -w(1)\frac{s(1-s)}{2} \cdot \left[\frac{1^{\frac{(1-s)}{2}}}{(1-s)/2} + \frac{1^{\left(\frac{s}{2}\right)}}{s/2} \right]$$

$$= -w(1)\frac{s(1-s)}{2} \cdot \left[\frac{2}{(1-s)} + \frac{2}{s} \right] = -w(1).$$

Back to Edwards, the full equation is now:

$$= \frac{1}{2} + w(1) + \left[\frac{s(1-s)}{2} \int_1^{\infty} w'(x)F(x)\, dx \right]$$

$$= \frac{1}{2} + w(1) + \left[\frac{s(1-s)}{2} \int_1^{\infty} w'(x) \left[\frac{x^{\frac{(1-s)}{2}}}{(1-s)/2} + \frac{x^{\left(\frac{s}{2}\right)}}{s/2} \right] dx \right]$$

$$= \frac{1}{2} + w(1) + \int_1^{\infty} w'(x) \left[sx^{\frac{(1-s)}{2}} + (1-s)x^{\left(\frac{s}{2}\right)} \right] dx$$

$$= \frac{1}{2} + w(1) + \int_1^{\infty} x^{\left(\frac{3}{2}\right)}w'(x) \left[sx^{\frac{(-s)}{2}-1} + (1-s)x^{\left(\frac{s-1}{2}\right)-1} \right] dx.$$

We now stop to evaluate the above integral (using the same version of Integration by Parts as above):

$$a(x) = x^{\left(\frac{3}{2}\right)}w'(x) \qquad b(x) = \left[sx^{\frac{(-s)}{2}-1} + (1-s)x^{\left(\frac{s-1}{2}\right)-1} \right]$$

$$B(x) = -2x^{\frac{(-s)}{2}} - 2x^{\frac{(s-1)}{2}}$$

Now substituting where needed:

$$\int_1^{\infty} x^{\left(\frac{3}{2}\right)}w'(x) \left[sx^{\frac{(-s)}{2}-1} + (1-s)x^{\left(\frac{s-1}{2}\right)-1} \right] dx = \int_1^{\infty} a(x)b(x)\, dx$$

$$= \int_1^{\infty} \frac{d}{dx}[a(x)B(x)]\, dx - \int_1^{\infty} \left[\frac{d}{dx}a(x) \right] B(x)\, dx$$

$$= \int_1^{\infty} \frac{d}{dx}\left[x^{\left(\frac{3}{2}\right)}w'(x)\left(-2x^{\frac{(-s)}{2}} - 2x^{\frac{(s-1)}{2}}\right) \right] dx - \int_1^{\infty} \left[\frac{d}{dx}a(x) \right] (B(x))\, dx.$$

But $w'(x) < w(x)$, so by the same logic we used above:

$$= \left[x^{\left(\frac{3}{2}\right)}w'(x)\left(-2x^{\frac{(-s)}{2}} - 2x^{\frac{(s-1)}{2}}\right) \right]_1^{\infty} - \int_1^{\infty} \left[\frac{d}{dx}a(x) \right] (B(x))\, dx$$

$$= -w'(x)[-2-2] - \int_1^{\infty} \left[\frac{d}{dx}a(x) \right] (B(x))\, dx$$

$$= 4w'(x) - \int_1^{\infty} \left[\frac{d}{dx}a(x) \right] (B(x))\, dx$$

$$= 4w'(x) - \int_1^\infty \left[\frac{d}{dx} x^{\left(\frac{3}{2}\right)} w'(x) \right] \left(-2x^{\frac{(-s)}{2}} - 2x^{\frac{(s-1)}{2}} \right) dx$$

$$= 4w'(x) + \int_1^\infty \left[\frac{d}{dx} x^{\left(\frac{3}{2}\right)} w'(x) \right] \left(2x^{\frac{(-s)}{2}} + 2x^{\frac{(s-1)}{2}} \right) dx.$$

Which brings us back to Edwards and the full equation:

$$= \frac{1}{2} + w(1) + 4w'(1) + \int_1^\infty \left[\frac{d}{dx} x^{\left(\frac{3}{2}\right)} w'(x) \right] \left(2x^{\frac{(-s)}{2}} + 2x^{\frac{(s-1)}{2}} \right) dx. \tag{8.2}$$

Next, we would like to deal separately with the first three terms. Below, we restate the inversion rule for $w(x)$ (lemma 7.3) and then differentiate both sides (the RHS uses the product rule):

$$2w(x) + 1 = x^{-1/2} \cdot [2w(1/x) + 1]$$

$$\frac{d}{dx}[2w(x) + 1] = \frac{d}{dx}\left[x^{-1/2} \cdot [2w(1/x) + 1] \right]$$

$$2w'(x) = \frac{d}{dx}\left[x^{-1/2} \right] \cdot [2w(1/x) + 1] + \left[x^{-1/2} \right] \cdot \frac{d}{dx}[2w(1/x) + 1].$$

But by the chain rule:

$$\frac{d}{dx}[2w(1/x) + 1] = 2w'(1/x) \cdot \frac{d}{dx}\left(\frac{1}{x} \right) = 2w'(1/x) \cdot -x^{-2}.$$

So, we have:

$$2w'(x) = \frac{d}{dx}\left[x^{-1/2} \right] \cdot [2w(1/x) + 1] + \left[x^{-1/2} \right] \cdot \frac{d}{dx}[2w(1/x) + 1]$$

$$= \left(-\frac{1}{2}\left[x^{-3/2} \right] \cdot [2w(1/x) + 1] \right) + \left[x^{-1/2} \right] \cdot 2w'(1/x) \cdot -x^{-2}.$$

Now let $x = 1$ to obtain:

$$2w'(1) = -\frac{1}{2} - w(1) - 2w'(1)$$

$$0 = \frac{1}{2} + w(1) + 4w'(1).$$

With the result above, we return to Edwards and equation (8.2) (dropping the first three terms):

$$\xi(s) = \int_1^\infty \left[\frac{d}{dx} x^{\left(\frac{3}{2}\right)} w'(x) \right] \left(2x^{\frac{(-s)}{2}} + 2x^{\frac{(s-1)}{2}} \right) dx.$$

Next, we simplify the right side (parenthetical) of the integral. We have:

$$2x^{\frac{(-s)}{2}} + 2x^{\left(\frac{s-1}{2}\right)} = 2\left[x^{\frac{(-s)}{2}} + x^{\left(\frac{s-1}{2}\right)} \right]$$

$$= 2\left[x^{-\left[\frac{1}{2}\left(s - \frac{1}{2}\right)\right] - \frac{1}{4}} + x^{\left[\frac{1}{2}\left(s - \frac{1}{2}\right)\right] - \frac{1}{4}} \right]$$

$$= 2x^{-1/4}\left[x^{-\left[\frac{1}{2}\left(s - \frac{1}{2}\right)\right]} + x^{\left[\frac{1}{2}\left(s - \frac{1}{2}\right)\right]} \right]$$

CHAPTER 8. COMPLETED ZETA FUNCTION

$$= 4x^{-1/4}\left[\frac{1}{2}\left(x^{-\left[\frac{1}{2}\left(s-\frac{1}{2}\right)\right]} + x^{\left[\frac{1}{2}\left(s-\frac{1}{2}\right)\right]}\right)\right]$$

$$= 4x^{-1/4}\left[\frac{1}{2}\left(e^{-\left[\frac{1}{2}\left(s-\frac{1}{2}\right)\right]log(x)} + e^{\left[\frac{1}{2}\left(s-\frac{1}{2}\right)\right]log(x)}\right)\right]$$

$$= 4x^{-1/4}cosh\left(\left[\frac{1}{2}\left(s-\frac{1}{2}\right)log(x)\right]\right).$$

Returning to the full equation, it is now in what Edwards calls the "final form":

$$\xi(s) = 4\int_1^\infty \left[\frac{d}{dx}x^{\left(\frac{3}{2}\right)}w'(x)\right]\left(x^{-1/4}\cosh\left(\left[\frac{1}{2}\left(s-\frac{1}{2}\right)log(x)\right]\right)\right)dx. \tag{8.3}$$

Now consider the *cosh* part of equation (8.3), which can be expanded into the usual power series. We have:

$$cosh(y) = \frac{1}{2}\left(e^y + e^{-y}\right) = \sum_0^\infty \frac{y^{2n}}{(2n)!}.$$

Therefore:

$$cosh\left(\left[\frac{1}{2}\left(s-\frac{1}{2}\right)log(x)\right]\right) = \sum_{n=0}^\infty \frac{\left[\frac{1}{2}\left(s-\frac{1}{2}\right)log(x)\right]^{2n}}{(2n)!}.$$

We now start with the "final form" of $\xi(s)$ and re-write it as follows:

$$\xi(s) = 4\int_1^\infty \left[\frac{d}{dx}x^{\left(\frac{3}{2}\right)}w'(x)\right]\left(x^{-1/4}cosh\left(\left[\frac{1}{2}\left(s-\frac{1}{2}\right)log(x)\right]\right)\right)dx$$

$$= 4\int_1^\infty \left[\frac{d}{dx}x^{\left(\frac{3}{2}\right)}w'(x)\right]\left(x^{-1/4}\sum_{n=0}^\infty \frac{\left[\frac{1}{2}\left(s-\frac{1}{2}\right)log(x)\right]^{2n}}{(2n)!}\right)dx$$

$$= 4\int_1^\infty \left[\frac{d}{dx}x^{\left(\frac{3}{2}\right)}w'(x)\right]\left(x^{-1/4}\sum_{n=0}^\infty \frac{\left(s-\frac{1}{2}\right)^{2n}\left(\frac{1}{2}log(x)\right)^{2n}}{(2n)!}\right)dx$$

$$= 4\int_1^\infty \left[\frac{d}{dx}x^{\left(\frac{3}{2}\right)}w'(x)\right]\left(x^{-1/4}\sum_{n=0}^\infty \left(s-\frac{1}{2}\right)^{2n}\frac{\left(\frac{1}{2}log(x)\right)^{2n}}{(2n)!}\right)dx.$$

Now recall our definition of a_{2n}

$$a_{2n} = 4\int_1^\infty \left[\frac{d}{dx}x^{\left(\frac{3}{2}\right)}w'(x)\right]\left(x^{-1/4}\frac{\left(\frac{1}{2}log(x)\right)^{2n}}{(2n)!}\right)dx.$$

We therefore have

$$\xi(s) = \sum_{n=0}^\infty a_{2n}\left(s-\frac{1}{2}\right)^{2n}. \qquad\qquad \square$$

NOTE: the fact that the coefficients a_{2n} are positive follows immediately from:

$$\frac{d}{dx}\left[x^{3/2}w'(x)\right] = \frac{d}{dx}\left(-\sum_{n=1}^\infty x^{3/2}n^2\pi e^{-n^2\pi x}\right) = \sum_{n=1}^\infty \left(n^4\pi^2 x - \frac{3}{2}n^2\pi\right)x^{1/2}e^{-n^2\pi x}$$

because this shows that the integrand (in the integral for a_{2n}) is positive for $x \geq 1$.

8.3   The Order of the Completed Zeta Function.

We start with two lemmas, and then show that the order of $\xi(s)$ is 1.

Lemma 8.1. *Let $z \in \mathbb{C}$. Then, for any $\epsilon > 0$ and $M > 1$, there is an $R > 1$ such that for $|z| > R$, we have $M \log |z| < |z|^\epsilon$ (or equally $\log |z| < |z|^\epsilon / M$).*

Proof. Fix ϵ and M. For our fixed ϵ, there is an $n \in \mathbb{N}$ such that $1/n < \epsilon$ and therefore $|z|^{1/n} < |z|^\epsilon$. Thus, it suffices to prove the lemma is true for any n. Fix n and set $R = e^{M^2 4n^4}$. Because $|z| > R$, we have $\log |z| > M^2 4n^4$, and $\sqrt{\log |z|} > M 2n^2$. Thus:

$$
\begin{aligned}
|z|^{1/n} = e^{\log|z|/n} &= \sum_{k=0}^{\infty} \frac{(\log |z|/n)^k}{k!} = \sum_{k=0}^{\infty} \frac{\left(\sqrt{\log|z|/n} \cdot \sqrt{\log|z|/n} \right)^k}{k!} \\
&\geq \sum_{k=0}^{\infty} \frac{\left(\sqrt{\log|z|/n} \cdot \sqrt{M2n^2/n} \right)^k}{k!} = \sum_{k=0}^{\infty} \frac{\left(\sqrt{\log|z|/n} \cdot \sqrt{M2n} \right)^k}{k!} = \sum_{k=0}^{\infty} \frac{\left(\sqrt{M2}\sqrt{\log|z|} \right)^k}{k!} \\
&> \frac{\left(\sqrt{M2}\sqrt{\log|z|} \right)^2}{2!} = M \log |z|. \qquad \square
\end{aligned}
$$

Lemma 8.2. *Let $R > 0$ and define the closed disk $\overline{D}_{s,R} = \{s \in \mathbb{C} : |s - 1/2| \leq R\}$. For sufficiently large R and $s \in \overline{D}_{s,R}$, we have $|\xi(s)| \leq R^R$.*

Proof. [19] From Theorem 8.1, we have:

$$
\xi(s) = \sum_{n=0}^{\infty} a_{2n} \left(s - \frac{1}{2} \right)^{2n}.
$$

where the coefficients a_{2n} are all positive. Thus, the largest value of $\xi(s)$ on $\overline{D}_{s,R}$ is at $s = 1/2 + R$, so we need only show that $\xi(1/2 + R) \leq R^R$ for sufficiently large R.

Recall that

$$
\xi(s) = \frac{s(s-1)}{2} \pi^{-s/2} \Gamma\left(\left(\frac{s}{2} \right) \right) \zeta(s),
$$

and that $\zeta(s)$ decreases to 1 as $s \to +\infty$.

Fix R and choose N such that $1/2 + R \leq 2N < 1/2 + R + 2$. We have (for some constant A):

$$
\xi(1/2 + R) \leq \xi(2N) = (N!)\pi^{-N}(2N - 1)\zeta(2N) \leq N^N \pi^{-0}(2N)\zeta(2) = A \cdot N^{N+1}
$$

But, since $N < R/2 + 2$, that gives $A \cdot N^{N+1} \leq A\,(R/2 + 2)^{R/2+3} < R^R$ for sufficiently large R. $\square$

The theorem immediately below assumes knowledge of section 11.3, where we discuss the definition of the order of an entire function.

Theorem 8.2. $\xi(s)$ *is of order 1.*

Proof #1 – Based on Ahlfors. [20] Because $\xi(s) = \xi(1-s)$, we need only estimate $|\xi(s)|$ for $Re(s) \geq 1/2$. From Stirling's formula [21] we have $\log|\Gamma(s/2)| \leq A|s|\log|s|$ (or equally, $|\Gamma(s/2)| \leq e^{A|s|\log|s|}$) for some constant A and large $|s|$ (importantly, this estimate is precise for real values of s). Applying lemma 8.1 we have $|\Gamma(s/2)| \leq e^{A|s|^{1+\epsilon}}$ for any $\epsilon > 0$ and sufficiently large $|s|$. The infinium of all such $1 + \epsilon$ is equal to 1. Therefore, if $|\zeta(s)|$ is sufficiently small when $Re(s) \geq 1/2$, we can conclude that the order of $\xi(s)$ is equal to 1.

We obtain the required estimate of $\zeta(s)$ from Theorem 5.2, because for $s = \sigma + it$, $\sigma \geq 1/2$ and $t \geq 2$ we have $|\zeta(s)| \leq 9t^{1/2}$. Thus, the factor $\zeta(s)$ does not increase the order of $\xi(s)$ beyond 1. $\qquad \square$

Proof #2 – Based on Edwards. We have from Lemma 8.2 that, for sufficiently large R, $|\xi(s)| \leq R^R$ throughout the disk $|s-1/2| \leq R$. But $R^R = e^{R\log R}$. By Lemma 8.1, for any $\epsilon > 0$, there is a sufficiently large R' such that $R'^{R'} = e^{R'\log R'} \leq e^{R'^{1+\epsilon}}$. That means the order of $\xi(s) \leq 1$. From Stirling's formula (see Proof #1), we have $\log|\Gamma(s/2)| \leq A|s|\log|s|$ for some constant A and large $|s|$, and this estimate is precise for real values of s. That means the order of $\xi(s) \geq 1$. Thus, the order of $\xi(s) = 1$. $\qquad \square$

8.4 The Product Formula for the Completed Zeta Function

We have shown that $\xi(z)$ is entire and has growth order 1. Now evaluate $\xi(1/2 + w)$. We have from Theorem 11.2 (where σ denotes the (non-zero) zeros of $\xi(1/2 + w)$):

$$\xi(1/2 + w) = e^{P(w)}w^m \prod_\sigma E_1(w/\sigma) = e^{P(w)}w^m \prod_\sigma \left(1 - \frac{w}{\sigma}\right)e^{w/\sigma}$$

where P is a polynomial of degree ≤ 1, and m is the order of the zero of f at $z = 0$.

We first simplify by noting that $\xi(1/2) \neq 0$ so $m = 0$ and $w^m = 1$. Also, if we group the σ and $-\sigma$ together, the exponential factors cancel. Finally, by Theorem 11.2, the polynomial $P(w)$ is of degree at most one, so can be restated as e^{A+Bw} for some A and B. We have simplified to

$$\xi(1/2 + w) = e^{A+Bw} \prod_\sigma \left(1 - \frac{w}{\sigma}\right).$$

Now note that the left-hand side and the product are both even functions. That means e^{A+Bw} must be an even function, which is only possible if $B = 0$, making e^{A+Bw} a constant, which we call c, leaving:

$$\xi(1/2 + w) = c \prod_\sigma \left(1 - \frac{w}{\sigma}\right).$$

Now undo the change of variables by setting $s = 1/2 + w$ and $\rho = 1/2 + \sigma$:

$$\xi(s) = c \prod_\rho \left(1 - \frac{s - 1/2}{\rho - 1/2}\right).$$

Next note that $\xi(0)$ is

$$\xi(0) = c \prod_\rho \left(1 - \frac{-1/2}{\rho - 1/2}\right),$$

so that dividing both sides by $\xi(0)$ gives

$$
\begin{aligned}
\frac{\xi(s)}{\xi(0)} &= \prod_\rho \left(1 - \frac{s - 1/2}{\rho - 1/2}\right)\left(1 - \frac{-1/2}{\rho - 1/2}\right)^{-1} \\
&= \prod_\rho \left(\frac{\rho - 1/2}{\rho - 1/2} - \frac{s - 1/2}{\rho - 1/2}\right)\left(\frac{\rho - 1/2}{\rho - 1/2} - \frac{-1/2}{\rho - 1/2}\right)^{-1} \\
&= \prod_\rho \left(\frac{\rho - s}{\rho - 1/2}\right)\left(\frac{\rho}{\rho - 1/2}\right)^{-1} \\
&= \prod_\rho \left(\frac{\rho - s}{\rho - 1/2}\right)\left(\frac{\rho - 1/2}{\rho}\right) = \prod_\rho \left(\frac{\rho - s}{\rho}\right) \\
&= \prod_\rho \left(1 - \frac{s}{\rho}\right).
\end{aligned}
$$

Finally, multiply both sides by $\xi(0)$ to obtain the final form of the product formula

$$
\xi(s) = \xi(0) \prod_\rho \left(1 - \frac{s}{\rho}\right) \qquad \text{with } \xi(0) = \tfrac{1}{2}, \tag{8.4}
$$

where ρ has its usual meaning (and ordering) as the roots of $\xi(s)$ (see page xi), and ρ and $(1 - \rho)$ are paired.[22]

The Density of the Zeros

9.1 Introduction

For $s \in \mathbb{C}, T > 0$, we define a rectangle $R(T)$ as $0 \leq Re(s) \leq 1$ and $0 \leq Im(s) \leq T$. We define $N(T)$ as the number of zeros of $\xi(s)$ inside the rectangle $R(T)$. In this chapter, our goal is to develop an estimate[23] for $N(T)$.

We will use the argument principle. Let $f(z)$ be a meromorphic function on an open set Ω and let C be a simple closed curve in Ω which avoids the zeros and poles of f. Then Cauchy's argument principle states:

$$\int_C \frac{f'(z)}{f(z)}\, dz = 2\pi i(N - P),$$

where N and P denote, respectively, the number of zeros and poles of $f(z)$ inside the contour C, with each zero and pole counted as many times as its multiplicity and order, respectively, indicate.

In our case, ξ is an entire function (no poles). We therefore have

$$N(T) = \frac{1}{2\pi i} \int_\gamma \frac{\xi'(z)}{\xi(z)}\, dz,$$

where the path γ represents a positive (counterclockwise) traversal of the boundary of $R(T)$.

To better understand the argument principle, consider the integral that results from a change of variable. Think of our current path γ as a parameterization $z = \gamma(t)$. Let $w = \xi(z)$, so that $w = \xi(\gamma(t))$ is another curve. Then $dw = \xi'(z)\, dz$ and our integral becomes

$$N(T) = \frac{1}{2\pi i} \int_{\xi(\gamma)} \frac{dw}{w}.$$

By assumption, γ does not go through any zeros of ξ, so $w = \xi(\gamma(t))$ is never zero and 1/w in the integral is not a problem. The properties of this logarithmic integral are well known. It is $2\pi i$ times the winding number of the closed path $\xi(\gamma)$ around the origin. Thus, our original integral is equal to the total change in the argument of $\xi(z)$ as z travels around γ, explaining the name of the theorem. In computing $N(T)$, then, our job is to determine the total change in argument of $\xi(z)$ for $z = \gamma(t)$.

9.2 Some Useful Lemmas

Lemma 9.1. *Let $R(T)$ and $N(T)$ be as defined above. Then, as $T \to \infty$, $N(T+1) - N(T) = \mathcal{O}(\log(T))$.*

Proof.[24] Let $\mathbf{n}(r)$ be the number of zeros of $\zeta(s)$ in the circle with center at $2+iT$ and radius r. Clearly, any circle with radius $r > \sqrt{5}$ covers all of the region included in $R(T+1) - R(T)$. We therefore have

$$N(T+1) - N(T) \le \mathbf{n}(\sqrt{5}).$$

By lemma 11.8 (Jensen's Formula) and lemma 11.9:

$$\int_0^3 \frac{\mathbf{n}(r)}{r}\, dr = \frac{1}{2\pi} \int_0^{2\pi} \log|\zeta(2+iT+3e^{i\theta})|\, d\theta - \log|\zeta(2+iT)|.$$

Let $\overline{D}_3(2+iT)$ be the closed disk of radius 3, centered at $2+iT$. For our fixed T choose A such that for $s \in \overline{D}_3(2+iT)$, we have $|\zeta(s)| < T^A$. That means $\log|\zeta(2+iT+3e^{i\theta})| < A\log(T)$. Thus,

$$\int_0^3 \frac{\mathbf{n}(r)}{r}\, dr < A\log T,$$

and we have our desired result:

$$A\log T > \int_0^3 \frac{\mathbf{n}(r)}{r}\, dr \ge \int_{\sqrt{5}}^3 \frac{\mathbf{n}(r)}{r}\, dr \ge \mathbf{n}(\sqrt{5}) \int_{\sqrt{5}}^3 \frac{dr}{r} = \left(\log 3 - \log\sqrt{5}\right)\mathbf{n}(\sqrt{5}). \qquad \square$$

It follows that $N(T+h) - N(T) = \mathcal{O}(\log(T))$ for any fixed $h > 0$. That also means that the multiplicity of multiple zeros in any such region is at most $\mathcal{O}(\log(T))$.

Lemma 9.2. *For $0 \le x \le 1/2$, we have: (i) $x/8 \le \log\sqrt{1+x} \le x$, and (ii) $\log\sqrt{1+x^2} \le x$.*

Proof. We have (because $x^2 < x$)

$$\left(1 + \frac{3x}{8}\right)^2 = 1 + \frac{6x}{8} + \frac{9x^2}{64} = 1 + \frac{48x}{64} + \frac{9x^2}{64} \le 1 + \frac{48x}{64} + \frac{9x}{64} = 1 + \frac{57x}{64} \le 1 + x.$$

So that

$$\log\left(1 + \frac{3x}{8}\right) = \log\sqrt{\left(1 + \frac{3x}{8}\right)^2} \le \log\sqrt{1+x}.$$

We now use the Taylor Series for the logarithm, valid for $|x| < 1$:

$$\log(1+x) = x - \frac{x^2}{2} + \frac{x^3}{3} - \frac{x^4}{4} + \frac{x^5}{5}\dots$$

Because $0 \le x \le 1/2$ (and therefore $x^n \ge x^{n+1}$ for $n \in \mathbb{N}$):

$$\frac{x}{2} = x - \frac{x}{2} \le x - \frac{x^2}{2} \le \log(1+x) \le x.$$

Putting it all together, for (i) we have:

$$\frac{x}{8} \le \frac{1}{2}\left(\frac{3x}{8}\right) \le \log\left(1 + \frac{3x}{8}\right) \le \log\sqrt{1+x} \le \log\sqrt{(1+x)^2} \le \log(1+x) \le x.$$

Note that (ii) follows immediately from (i) because $x^2 \le x$. $\qquad \square$

CHAPTER 9. THE DENSITY OF THE ZEROS

Lemma 9.3. *Let $z = x + iT$ with $0 \leq x \leq 1/2$, $T > 2$, and with T assumed increasing:*

$$\log(x + iT) = i\left[\pi/2 - \arctan(x/T)\right] + \log(T) + \mathcal{O}(1/T) \tag{9.1}$$

$$= i\left[\pi/2 - \mathcal{O}(1/T)\right] \quad + \log(T) + \mathcal{O}(1/T) \tag{9.2}$$

$$\log(-x + iT) = i\left[\pi/2 + \arctan(x/T)\right] + \log(T) + \mathcal{O}(1/T) \tag{9.3}$$

$$= i\left[\pi/2 + \mathcal{O}(1/T)\right] \quad + \log(T) + \mathcal{O}(1/T) \tag{9.4}$$

Proof. We note that we are in the first quadrant (the coordinates of $x + iT$ are both positive), so we can unambiguously use $\arctan(y/x)$. We have

$$\log(x + iT) = i\arctan(T/x) + \log\sqrt{x^2 + T^2}$$

$$= i\arctan(T/x) + \log\sqrt{T^2} + \log\sqrt{x^2 + T^2} - \log\sqrt{T^2}$$

$$= i\arctan(T/x) + \log(T) + \log\sqrt{\frac{x^2 + T^2}{T^2}}$$

$$= i\arctan(T/x) + \log(T) + \log\sqrt{\left(\frac{x}{T}\right)^2 + 1} \tag{9.5}$$

For increasing T, it is easily verified that $\arctan(T/x)$ rapidly approaches $\pi/2$ from below. Using the formula $\arctan(x) = \pi/2 - \arctan(1/x)$, we have $\arctan(T/x) = \pi/2 - \arctan(x/T)$. From the arctan Taylor Series (just below), $\arctan(x/T) \leq x/T$ thus $\leq 1/2T$, so rapidly approaches 0.

$$\arctan(x) = \sum_{n=0}^{\infty} \frac{(-1)^n}{2n+1} x^{2n+1} = x - \frac{x^3}{3} + \frac{x^5}{5} - \frac{x^7}{7} + \dots \quad \text{for } x < 1.$$

Regarding our second logarithm in equation (9.5) above, from lemma 9.2 we have

$$\log\sqrt{\left(\frac{x}{T}\right)^2 + 1} \leq \frac{x}{T} \leq \frac{1}{2T} = \mathcal{O}(1/T).$$

Putting it all together, we have

$$\log(x + iT) = i\left[\pi/2 - \arctan(x/T)\right] + \log(T) + \mathcal{O}(1/T)$$

$$= i\left[\pi/2 - \mathcal{O}(1/T)\right] \quad + \log(T) + \mathcal{O}(1/T).$$

For $\log(-x + iT)$ the stated results follow immediately by noting that $\arctan(-x) = -\arctan(x)$. $\quad\square$

9.3 Two Important Lemmas

Lemma 9.4. *Let $f(s = x + iy)$ be a holomorphic function on a vertical strip $-1 \leq x \leq 4$, except possibly for a simple pole at $s = 1$. Suppose $f(s)$ is real for real s (i.e., $f(\bar{s}) = \overline{f(s)}$). Assume a lower bound $|f(2 + iy)| \geq m > 0$, and a family of upper bounds*

$$|f(x + iy)| \leq M(T) \quad \text{(for } 1/4 \leq x \leq 4 \text{ and } 1 \leq y \leq T\text{)}.$$

Then, for T not a vertical coordinate of a zero of f, there is the following upper bound for the change in argument from $2 + iy$ to $1/2 + iy$:

$$|\arg f(1/2 + iy) - \arg f(2 + iy)| \leq \frac{\pi}{\log\left[(1.75)/(1.5)\right]} \cdot \left[\log(M(T+2)) + \log(1/m)\right] + \pi.$$

Proof. [25] Let q be the number of times $Re(f(x + iT))$ changes sign along the path $2 + iT$ to $1/2 + iT$. The points of change of sign divide the path into $q + 1$ sub-paths; on each sub-path, either $Re(f) \geq 0$ or $Re(f) \leq 0$. Clearly, arg f cannot change by more than π within any one sub-path, so that the total change of arg f along the full path is $\leq (q + 1) \cdot \pi$.

Our goal is to obtain an upper bound for q. To assist in that, we define

$$g(z) = 1/2 [f(z + iT) + f(z - iT)], \quad z \in \mathbb{C}.$$

Like $f(s)$, we have that $g(z)$ is holomorphic on the vertical strip $\{z \in \mathbb{C} : -1 \leq Re(z) \leq 4\}$, with a possible pole at $s = 1$.

Now define $L = \{z \in \mathbb{C} : Im(z) = 0, 1/2 \leq Re(z) \leq 2\}$ and $\overline{D}_2(r) = \{z \in \mathbb{C} : |z - 2| \leq r\}$, so that L is a line segment on the real line and $\overline{D}_2(r)$ is a closed disk centered at 2 with radius r. Note that $L \in \overline{D}_2(1.5)$.

For $z_L \in L$, we next show that $g(z_L) = 0$ *if and only if* $Re(f(z_L + iT)) = 0$. By assumption, $f(\overline{s}) = \overline{f(s)}$, and by definition $Re(\overline{f(s)}) = Re(f(s))$. Thus $Re(f(s)) = Re(f(\overline{s}))$. If $g(z_L) = 0$, then $Re(f(z_L + iT)) = Re(f(z_L - iT))$, and therefore $Re(f(z_L + iT)) = 0$. If $Re(f(z_L + iT)) = 0$, then $f(z_L + iT) = Im(f(z_L + iT))$ and $f(z_L - iT) = -Im(f(z_L + iT))$ so that $g(z_L) = 0$.

Note that the count q is the number of zeros of $g(z_L)$. Because $x_L \in L \in \overline{D}_2(1.5)$, the count q is bounded by the number of zeros of $g(z)$ for $z \in \overline{D}_2(1.5)$. We now obtain that bound.

Let $\mathbf{n}(r)$ be the number of zeros of $g(z)$ for $z \in \overline{D}_2(r)$. Because $q \leq \mathbf{n}(1.5)$, we have:

$$\int_0^{1.75} \frac{\mathbf{n}(r)}{r} \, dr \geq \int_{1.5}^{1.75} \frac{\mathbf{n}(r)}{r} \, dr \geq \int_{1.5}^{1.75} \frac{\mathbf{n}(1.5)}{r} \, dr = \mathbf{n}(1.5) \int_{1.5}^{1.75} \frac{1}{r} \, dr$$

$$= \mathbf{n}(1.5) \left[\log(1.75) - \log(1.5) \right] = \mathbf{n}(1.5) \cdot \log \frac{1.75}{1.5} \geq q \cdot \log \frac{1.75}{1.5}.$$

By lemma 11.8 (Jensen's Formula) and lemma 11.9,

$$\int_0^{1.75} \frac{\mathbf{n}(r)}{r} \, dr = \frac{1}{2\pi} \int_0^{2\pi} \log |g(2 + (1.75)e^{i\theta})| \, d\theta - \log |g(2)| \leq \log [M(T + 2)] + \log 1/m$$

Putting the above results together, we have proved the lemma

$$q \cdot \log \frac{1.75}{1.5} \leq \int_0^{1.75} \frac{\mathbf{n}(r)}{r} \, dr \leq \log [M(T + 2)] + \log 1/m$$

$$q \leq \frac{\log [M(T + 2)] + \log 1/m}{\log [1.75/1.5]}. \qquad \square$$

Lemma 9.5. *For $T > 2$ (and assumed increasing), we compute the net change in the imaginary part of certain logarithms along a given path, as follows:*

$$(i) \qquad Im \left(\log \left(\frac{s(s - 1)}{2} \right) \Big|_2^{1/2+iT} \right) = \pi$$

$$(ii) \qquad Im \left(\log \left(\pi^{\frac{-s}{2}} \right) \Big|_2^{1/2+iT} \right) = \frac{-T}{2} \log \pi$$

$$(iii) \qquad Im \left(\log \left(\Gamma \left(\frac{s}{2} \right) \right) \Big|_2^{1/2+iT} \right) = \frac{T}{2} \log \left(\frac{T}{2} \right) - \frac{T}{2} - \frac{\pi}{8} + \mathcal{O}(T^{-1}).$$

$$(iv) \qquad Im \left(\log \left(\zeta(s) \right) \Big|_2^{1/2+iT} \right) \leq \mathcal{O}(\log(T)).$$

Proof – Equation (i). We have

$$\log\left(\frac{s(s-1)}{2}\right)\Big|_2^{1/2+iT} = \log(s) + \log(s-1) - \log(2)\Big|_2^{1/2+iT}$$

$$= \left[\log(1/2+iT) + \log(-1/2+iT) - \log(2)\right] - \left[\log(2) + \log(1) - \log(2)\right]$$

$$= \log(1/2+iT) + \log(-1/2+iT) - \log(2)$$

$$= i\pi + 2\log\left(\frac{T}{2}\right) + \mathcal{O}(1/T) - \log(2) \qquad \text{(see lemma 9.3).}$$

The imaginary part is as stated in Equation (i). $\qquad\square$

Proof – Equation (ii). We have

$$\log\left(\pi^{\frac{-s}{2}}\right)\Big|_2^{1/2+iT} = -\frac{s}{2}\log\pi\Big|_2^{1/2+iT} = \frac{-(1/2+iT)}{2}\log\pi - \frac{2}{2}\log\pi = \frac{-(1/2+iT)}{2}\log\pi - \log\pi$$

$$= \frac{-iT}{2}\log\pi - 1.25\log\pi.$$

The imaginary part is as stated in Equation (ii). $\qquad\square$

Proof – Equation (iii). We first note the following formula[26]

$$\log\Gamma(s) = \left(s - \frac{1}{2}\right)\log(s) - s + \frac{1}{2}\log 2\pi + \mathcal{O}(s^{-1}) \quad \text{for } Re(s) > 0 \text{ as } |s| \to \infty$$

so we have

$$\log\Gamma(s/2)\Big|_2^{1/2+iT} = \log\Gamma\left(\frac{1/2+iT}{2}\right) - \log\Gamma(2/2) = \log\Gamma\left(\frac{1/2+iT}{2}\right) - 0$$

$$Im\left(\log\Gamma(s/2)\Big|_2^{1/2+iT}\right) = Im\left(\left[\frac{1/2+iT}{2} - \frac{1}{2}\right]\log\left(\frac{1/2+iT}{2}\right) - \left(\frac{1/2+iT}{2}\right) + \frac{1}{2}\log 2\pi + \mathcal{O}(T^{-1})\right)$$

$$= Im\left(\left[\frac{1/2+iT}{2} - \frac{1}{2}\right]\log\left(\frac{1/2+iT}{2}\right) - \left(\frac{1/2+iT}{2}\right)\right)$$

$$= Im\left(\left[\frac{iT}{2} - \frac{1}{4}\right]\log\left(\frac{1/2+iT}{2}\right) - \frac{1}{4} - \frac{iT}{2}\right)$$

$$= Im\left(\left[\frac{iT}{2} - \frac{1}{4}\right]\log\left(\frac{1/2+iT}{2}\right) - \frac{iT}{2}\right)$$

$$= Im\left(\left[\frac{iT}{2} - \frac{1}{4}\right]\left(i\left[\pi/2 - \mathcal{O}(T^{-1})\right] + \log\left(\frac{T}{2}\right) + \mathcal{O}(T^{-1})\right) - \frac{iT}{2}\right)$$

$$= Im\left(\frac{iT}{2}\left(i\left[\pi/2 - \mathcal{O}(T^{-1})\right] + \log\left(\frac{T}{2}\right) + \mathcal{O}(T^{-1})\right)\right)$$

$$\quad - Im\left(\frac{1}{4}\left(i\left[\pi/2 - \mathcal{O}(T^{-1})\right] + \log\left(\frac{T}{2}\right) + \mathcal{O}(T^{-1})\right)\right) - Im\left(\frac{iT}{2}\right)$$

$$= Im\left(\frac{iT}{2}\left(\log\left(\frac{T}{2}\right) + \mathcal{O}(T^{-1})\right)\right) - Im\left(\frac{i\pi}{8} + \frac{i}{4}\mathcal{O}(T^{-1})\right) - Im\left(\frac{iT}{2}\right)$$

$$= \frac{T}{2}\log\left(\frac{T}{2}\right) + \frac{T}{2}\mathcal{O}(T^{-1}) - \frac{\pi}{8} + \frac{1}{4}\mathcal{O}(T^{-1}) - \frac{T}{2}$$

In the factor $T/2 \cdot \mathcal{O}(T^{-1})$, we have from Lemma 9.3 that $\mathcal{O}(T^{-1}) \leq 1/2T$, so $T/2 \cdot \mathcal{O}(T^{-1}) \leq 1/4$ for all T, so can be disregarded for increasing T. Thus, the imaginary part is as stated in Equation (iii). $\qquad\square$

Proof – Equation (iv). In this last case, we analyze separately the two sections of the path.

$$Im\left(\log\left(\zeta(s)\right)\Big|_{2}^{1/2+iT}\right) = Im\left(\log\left(\zeta(s)\right)\Big|_{2}^{2+iT}\right) + Im\left(\log\left(\zeta(s)\right)\Big|_{2+iT}^{1/2+iT}\right)$$

For the path from 2 to $2 + iT$, we have $\arg f(2) = 0$. Also, $Re(f(s)) \neq 0$ along the path (because $Re(s) > 1$). We must therefore have $\arg f(s) = \arctan\left(Im(f(s))/Re(f(s))\right) < \pi/2 = \mathcal{O}(1)$.

For the path from $2 + iT$ to $1/2 + iT$, we apply lemma 9.4, with $f(s) = \zeta(s)$, $M(T) = 15T^{3/4}$ (from Theorem 5.2), and $m = \pi^2/6$. Thus

$$Im\left(\log\left(\zeta(s)\right)\Big|_{2+iT}^{1/2+iT}\right) \leq \frac{\pi}{\log\left[(1.75)/(1.5)\right]} \cdot \left[\log(15(T+2)^{3/4}) + \log(6/\pi^2)\right] + \pi$$

$$\leq \mathcal{O}(\log T).$$

Combining the two paths, we have

$$Im\left(\log\left(\zeta(s)\right)\Big|_{2}^{1/2+iT}\right) = \mathcal{O}(1) + \mathcal{O}(\log T) = \mathcal{O}(\log T).$$

To this point, we have assumed that T is not the ordinate of a zero of $\zeta(s)$. It follows from lemma 9.1 that $N(T+h) - N(T) = O(\log(T))$ for any fixed $h > 0$. So, if needed, we can increase T by a very small amount without changing the result. $\qquad\square$

9.4   The Riemann-Von Mangoldt Formula for $N(T)$

Theorem 9.1. *We recall $\xi(s)$, the Completed Zeta Function, $\xi(s) = \dfrac{s(s-1)}{2}\pi^{\frac{-s}{2}}\Gamma\left(\dfrac{s}{2}\right)\zeta(s)$.*

Let $s = \sigma + it$ and let $N(T)$ denote the number of zeros of $\xi(s)$ in $0 \leq \sigma \leq 1, 0 \leq t \leq T$. Then

$$N(T) = \frac{T}{2\pi}\log\left(\frac{T}{2\pi}\right) - \frac{T}{2\pi} + \mathcal{O}(\log(T)).$$

Proof. For the eight points $p_1, p_2, ..., p_8$ in the complex plane define the contour R to be the closed path

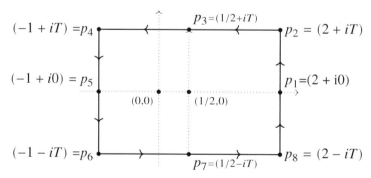

Figure 9.1: Contour Used to Compute N(T)

$p_1 \to p_2 \to p_3 \to p_4 \to p_5 \to p_6 \to p_7 \to p_8 \to p_1$. By definition, the part of the rectangle R in the upper half plane contains exactly the zeros counted by $N(T)$. In addition, because $\xi(\bar{s}) = \overline{\xi(s)}$, there are exactly the same number of zeros in R in the lower half plane. Using Cauchy's argument principle, that means

$$2N(T) = \frac{1}{2\pi i} \int_R \frac{\xi'(z)}{\xi(z)} dx.$$

Now define a path $\gamma = p_1 \to p_2 \to p_3$ (from $2 \to (2+iT) \to (1/2+iT)$). Using the functional equation $\xi(s) = \xi(1-s)$ and the symmetry $\xi(\bar{s}) = \overline{\xi(s)}$ we can integrate along γ (1/4 of the path R) and have

$$\frac{2N(T)}{4} = \frac{N(T)}{2} = \frac{1}{2\pi i} \int_\gamma \frac{\xi'(z)}{\xi(z)} dx$$

so that

$$N(T) = \frac{1}{\pi i} \int_\gamma \frac{\xi'(z)}{\xi(z)} dx.$$

Our plan, therefore, is to calculate $N(T)$ by integrating along the path γ. Specifically, we evaluate the net change in the imaginary part of the above integral over γ. We have

$$\int_\gamma \frac{\xi'(z)}{\xi(z)} dx = \log(\xi(s)) \Big|_2^{1/2+iT}$$

so we need to calculate

$$Im\left(\log(\xi(s))\Big|_2^{1/2+iT}\right) = Im\left(\left[\log(\frac{s(s-1)}{2}) + \log(\pi^{\frac{-s}{2}}) + \log(\Gamma\left(\frac{s}{2}\right)) + \log(\zeta(s))\log(\xi(s))\right]\Big|_2^{1/2+iT}\right)$$

But we have done precisely that in Lemma 9.5 and have

$$Im\left(\log(\xi(s))\Big|_2^{1/2+iT}\right) = \pi + \frac{-T}{2}\log\pi + \frac{T}{2}\log\left(\frac{T}{2}\right) - \frac{T}{2} - \frac{\pi}{8} + \mathcal{O}(T^{-1}) + \mathcal{O}(\log(T))$$

$$= \frac{-T}{2}\log\pi + \frac{T}{2}\log\left(\frac{T}{2}\right) - \frac{T}{2} - \frac{\pi}{8} + \pi + bigO(T^{-1}) + \mathcal{O}(\log(T))$$

$$= \frac{-T}{2}\log\pi + \frac{T}{2}\log\left(\frac{T}{2}\right) - \frac{T}{2} + \mathcal{O}(\log(T))$$

Therefore

$$N(T) = \frac{1}{\pi}Im\left(\log(\xi(s))\Big|_2^{1/2+iT}\right) = \frac{1}{\pi}\left(\frac{-T}{2}\log\pi + \frac{T}{2}\log\left(\frac{T}{2}\right) - \frac{T}{2} + \mathcal{O}(\log(T))\right)$$

$$= \frac{-T}{2\pi}\log\pi + \frac{T}{2\pi}\log\left(\frac{T}{2}\right) - \frac{T}{2\pi} + \mathcal{O}(\log(T))$$

$$= \frac{T}{2\pi}\left[\log\left(\frac{T}{2}\right) - \log\pi\right] - \frac{T}{2\pi} + \mathcal{O}(\log(T))$$

$$= \frac{T}{2\pi}\log\left(\frac{T}{2\pi}\right) - \frac{T}{2\pi} + \mathcal{O}(\log(T)) \qquad \square$$

Lemma 9.6. *Let $R(T)$ and $N(T)$ be as defined above. Then, for sufficiently large T, $N(T+1) - N(T) \leq 2\log T$.*

Proof. This is a refinement of lemma 9.1. Four years after Bäcklund's 1914 proof of Theorem 9.1, he sharpened his estimate and obtained [5], for all $T \geq 2$

$$\left| N(T) - \frac{T}{2\pi} \log\left(\frac{T}{2\pi}\right) - \frac{T}{2\pi} + \frac{7}{8} \right| < 0.137 \log T + 0.443 \log\log T + 4.35.$$

We will use that estimate in our proof. For another approach see Edwards [13, pp. 56-58].

Before we start, we develop some useful estimates. First, let $A = (T+1)/T$. Note that $A = 1 + 1/T$, so that (using the Taylor series) $\log A < 1/T$. Next note that $\log\log 2 < 0$ and $1/\log 2 < 1.5$.

Finally, we will use that $\log\log T < \log\log AT < \log\log 2T$ and develop an estimate for $\log\log 2T$. For our fixed T, fix $k \in \mathbb{N}$ such that $2^k \leq T < 2^{k+1}$. We have

$$\frac{\log\log 2T}{\log T} \leq \frac{\log\log 2^{k+2}}{\log T} \leq \frac{\log[(k+2)\log 2]}{k\log 2} < \frac{\log(k+2)}{k\log 2} < \frac{1.5\log(k+2)}{k}.$$

Using the Bäcklund estimate, we compute the maximum number of roots ρ in the interval $(T, T+1]$. To do so, for our fixed T, we compute the maximum value of $N(T+1)$ and subtract the minimum value of $N(T)$:

$$
\begin{aligned}
&= \left(\frac{T+1}{2\pi}\log\left(\frac{T+1}{2\pi}\right) - \frac{T+1}{2\pi} + \frac{7}{8} + [0.137\log(T+1) + 0.443\log\log(T+1) + 4.35]\right) \\
&\quad - \left(\frac{T}{2\pi}\log\left(\frac{T}{2\pi}\right) - \frac{T}{2\pi} + \frac{7}{8} - [0.137\log(T) + 0.443\log\log(T) + 4.35]\right) \\
&= \left[\frac{T+1}{2\pi}\log\left(\frac{AT}{2\pi}\right) - \frac{T}{2\pi}\log\left(\frac{T}{2\pi}\right)\right] - \left[\frac{T+1}{2\pi} - \frac{T}{2\pi}\right] + \left[\frac{7}{8} - \frac{7}{8}\right] \\
&\quad + 0.137\,(\log AT + \log T) + 0.443\,(\log\log AT + \log\log T) + 8.70 \\
&= \frac{1}{2\pi}\log\left(\frac{AT}{2\pi}\right) + \frac{T}{2\pi}\left[\log\left(\frac{AT}{2\pi}\right) - \log\left(\frac{T}{2\pi}\right)\right] - \frac{1}{2\pi} + 0 \\
&\quad + 0.137\,(\log A + \log T + \log T) + 0.443\,[\log(\log AT) + \log\log T] + 8.70 \\
&= \frac{1}{2\pi}\,(\log A + \log T - \log 2\pi) + \frac{T}{2\pi}\,(\log A + \log T - \log 2\pi - \log T + \log 2\pi) - \frac{1}{2\pi} \\
&\quad + 0.137\log A + 0.274\log T + 0.443\,[\log(\log AT) + \log\log T] + 8.70 \\
&= \frac{(T+1)\log A}{2\pi} + \frac{\log T}{2\pi} - \frac{1+\log 2\pi}{2\pi} \\
&\quad + 0.137\log A + 0.274\log T + 0.443\,[\log(\log AT) + \log\log T] + 8.70.
\end{aligned}
$$

We can now insert our estimates. We will discard negative constant terms, rearrange constant terms and assume that our fixed $T \geq 2^{14} = 16384$.

$$
\begin{aligned}
&\leq \frac{(T+1)}{T2\pi} + \frac{\log T}{2\pi} + \frac{0.137}{2^{14}} + 0.274\log T + \log T \cdot 0.443\left[2 \cdot 1.5 \cdot \frac{\log(k+2)}{k}\right] + 8.70 \\
&\leq \frac{\log T}{2\pi} + 0.274\log T + \log T \cdot 1.33\left[\frac{\log(k+2)}{k}\right] + \left(\frac{1}{2\pi} + \frac{1}{2^{14} \cdot 2\pi} + \frac{0.137}{2^{14}} + 8.70\right) \\
&\leq 0.5\log T + \log T \cdot 1.33\left[\frac{\log(k+2)}{k}\right] + 9 \\
&\leq 0.5\log T + 0.2\log T + \log T \boxed{\leq 1.7\log T}.
\end{aligned}
$$

$\square$

9.5 Using The Riemann-Von Mangoldt Formula

Inequalities that include sums over the roots of $\zeta(s)$ appear in multiple places in the theory of the zeta function. It is often critical to prove such sums are convergent. The lemmas below demonstrate the usefulness of the Riemann-Von Mangoldt Formula in those proofs. These lemmas will be used later in this book.

Lemma 9.7. *Let $\epsilon > 0$ be given, and let ρ have its usual meaning (and ordering) as the roots of $\xi(s)$ (see page xi). Let ρ_n be the nth ordered root. Then the following series converges:*

$$\sum_\rho \frac{1}{|\rho|^{1+\epsilon}}.$$

Proof. Create an ordered sequence of positive real numbers $R_1, R_2, R_3 ... R_n ...$, where R_n is defined as that number needed to make

$$R_n \log(R_n) = n \quad \text{for } n \in \mathbb{N}.$$

Now define the closed disk $\overline{D}_R = \{z \in \mathbb{C} : |z - \frac{1}{2}| \le R\}$. It is a simple application of Theorem 9.1 to see that there is a $K_1 > 0$ such that for $n > K_1$ the roots $\rho_n, \rho_{n+1} ...$ are not inside the disk $\overline{D}_{R_n}$.

Note also, there is a $K_2 > 0$ such that, for $n > K_2$ we have $(\log n)^2 < n^{\epsilon/2}$. This follows from lemma 8.1 (after taking the square root of both sides). Let $K = \max(K_1, K_2)$. Then

$$\sum_{n>K} \frac{1}{|\rho_n|^{1+\epsilon}} \le \sum_{n>K} \frac{1}{R_n^{1+\epsilon}} = \sum_{n>K} \frac{\log R_n^{1+\epsilon}}{n^{1+\epsilon}} = \sum_{n>K} \frac{1}{n^{1+\epsilon/2}} \frac{\log R_n^{1+\epsilon}}{n^{\epsilon/2}}$$

We have $\log n = \log R_n + \log \log R_n > \log R_n$ so that $(\log R_n)^{1+\epsilon} < (\log n)^2$. We previously established that, for our given K, $(\log n)^2 < n^{\epsilon/2}$. We therefore have

$$\le \sum_{n>K} \frac{1}{n^{1+\epsilon/2}} < \infty. \qquad \square$$

Lemma 9.8. *Let $s \in \mathbb{C}$, let $K > 0$ be fixed, and let ρ have its usual meaning (and ordering) as the roots of $\xi(s)$ (see page xi). Then the following series converges uniformly on $|s| < K$:*

$$\sum_\rho \frac{1}{(s - \rho)}.$$

Proof. We pair ρ and $(1 - \rho)$, let $x = (s - 1/2)$ and let $y = (\rho - 1/2)$. We then have

$$\left| \frac{1}{s - \rho} + \frac{1}{s - (1 - \rho)} \right| = \left| \frac{1}{x - y} + \frac{1}{x + y} \right| = \left| \frac{x + y}{(x + y)(x - y)} + \frac{x - y}{(x - y)(x + y)} \right|$$

$$= \left| \frac{2x}{x^2 - y^2} \right| \le \frac{2|x|}{||x^2| - |y^2||} = \frac{2|x|}{||y^2| - |x^2||}$$

We disregard the finite number of ρ where $|(\rho - 1/2)^2)| \le (K + 1)^2$

$$\le \frac{2(K + 1)}{|(\rho - 1/2)^2)| - (K + 1)^2} = \mathcal{O}\left(\frac{1}{|\rho|^2} \right),$$

for sufficiently large ρ (and our fixed K), which converges by lemma 9.7. $\qquad \square$

Chapter 10

Riemann's Explicit Formula

10.1 Introduction

The primary goal and main result of Riemann's paper is obtaining an *explicit formula* for the prime counting function, $\pi(x)$. The bulk of the work in getting to that goal is the development of a function which (following Edwards) we will call $J(x)$ – Riemann used the more generic $F(x)$. It is the relationship between $J(x)$ and both $\pi(x)$ and $\zeta(s)$ that leads to the *explicit formula*. Riemann shows:

$$J(x) = li(x) - \sum_{Im(\rho)>0} \left[li(x^\rho) + li\left(x^{1-\rho}\right) \right] + \int_x^\infty \frac{dt}{t\,(t^2 - 1)\,log(t)} + log(\xi(0)) \qquad \text{for } x > 1,$$

where $li(x)$ is the *logarithmic integral*, defined [27] for all positive real numbers $x \neq 1$ by

$$li(x) = \int_0^x \frac{dt}{\log t} \quad \text{or, for } x > 1 \quad \lim_{\epsilon \to 0+} \left[\int_0^{1-\epsilon} \frac{dt}{\log t} + \int_{1+\epsilon}^x \frac{dt}{\log t} \right].$$

Because $\pi(x)$ can be expressed in terms of $J(x)$, the $J(x)$ function provides a formula for $\pi(x)$. That is not to say you could, after Riemann's paper, use the formula to actually compute $\pi(x)$. However, the $J(x)$ function and Riemann's other results vastly improve our knowledge of $\pi(x)$.

There were some holes in Riemann's derivation of the *explicit formula*. The first complete proof was published by von Mangoldt in 1895 [26]. We will discuss (and supplement) Riemann's proof in this chapter.[28] We discuss von Mangoldt's proof in Chapter 13.

10.2 Some Preliminaries

10.2.1 The Relationship Between $\pi(x)$ and $J(x)$

Riemann starts with a slightly modified version of $\pi(x)$, the prime counting function:

$$\pi_0(x) = \frac{1}{2} \left[\sum_{p<x} 1 + \sum_{p \leq x} 1 \right] \qquad \text{(where } p \text{ is an ordered list of primes).}$$

The first sum counts the number of primes *less than* x, while the second sum counts the number of primes *less than or equal to* x. So, $\pi_0(x)$ almost always equals $\pi(x)$. However, if x is exactly a prime, then $\pi_0(x) = \pi(x) - 1/2$. From here on, when we write $\pi(x)$ we will mean $\pi_0(x)$ unless then context clearly indicates otherwise.

Riemann's next jump function is $J(x)$. The amount of the (full) jump is 1 for every prime, 1/2 for every prime square, 1/3 for every prime cube, etc. It has the same edge condition as $\pi_0(x)$: half of the jump amount happens *exactly at* the jump, while the other half happens *immediate after* the jump:

$$J(x) = \frac{1}{2}\left[\sum_{p^n<x}\frac{1}{n} + \sum_{p^n\leq x}\frac{1}{n}\right] \qquad \text{(where } p \text{ is an ordered list of primes).}$$

It is clear that $J(x) < x$ because it increases by a jump of ≤ 1 at each integer. Because the number of prime squares less than x is equal to the number of primes less than $x^{1/2}$ (i.e., $\pi(x^{1/2})$), and so on for higher prime powers, we see that $J(x)$ and $\pi(x)$ are related by

$$J(x) = \pi(x) + \frac{1}{2}\pi(x^{1/2}) + \frac{1}{3}\pi(x^{1/3}) + \dots \qquad x > 0. \tag{10.1}$$

Observe that, for any given $x > 0$, there is an N such that $x^{1/n} < 2$ for $n > N$. Thus, for any given x, the series has only a finite number of non-zero terms.

Now we would like to reverse the above equation and define $\pi(x)$ in terms of $J(x)$. We define the Möbius function $\mu(n)$:

$$\mu(n) = \begin{cases} 1 & \text{if } n = 1, \\ (-1)^k & \text{if n is the product of } k \text{ distinct primes,} \\ 0 & \text{otherwise (i.e., } n \text{ divisible by a prime square).} \end{cases} \tag{10.2}$$

Using Möbius inversion[29] on equation (10.1), we have

$$\pi(x) = \sum_{n=1}^{\infty} \frac{\mu(n)}{n} J(x^{1/n}). \tag{10.3}$$

Again, for any given x, the series has only a finite number of non-zero terms.

10.2.2   The Relationship Between $\zeta(s)$ and $J(x)$

To develop the relationship between $J(x)$ and $\zeta(s)$, we start with the product formula for $\zeta(s)$ for $Re(s) > 1$

$$\zeta(s) = \prod_p \frac{1}{1 - p^{-s}} \qquad \text{(where } p \text{ is an ordered list of primes).}$$

Now use the logarithm and the Taylor series for $\log(1 - z)$ with $|z| < 1$

$$\begin{aligned} \log \zeta(s) &= \sum_p -\log(1 - p^{-s}) \\ &= \sum_p \left(p^{-s} + 2p^{-2s} + 3p^{-3s} + \dots\right) \\ &= \sum_p \left(\sum_n \frac{1}{n} \cdot p^{-ns}\right). \end{aligned}$$

Next, noting that

$$p^{-ns} = s \int_{p^n}^{\infty} x^{-s-1}\, dx,$$

we have

$$\log \zeta(s) = s \sum_p \left(\sum_n \frac{1}{n} \int_{p^n}^{\infty} x^{-s-1} \, dx \right).$$

The last is absolutely convergent for $Re(s) > 1$, allowing us to move the sums inside the integral

$$\log \zeta(s) = s \int_0^{\infty} \sum_{p^n < x} \frac{1}{n} x^{-s-1} \, dx$$
$$= s \int_0^{\infty} J(x) x^{-s-1} \, dx.$$

Having defined $\log \zeta(s)$ in terms of $J(x)$, we would like to do the reverse and define $J(x)$ in terms of $\log \zeta(s)$. We have for $Re(s) > 1$:

$$\frac{\log \zeta(s)}{s} = \int_0^{\infty} J(x) x^{-s-1} \, dx. \tag{10.4}$$

To see that $J(x) x^{-s-1}$ is integrable (and finite) on $[0, \infty)$, we have previously shown that $J(x) < x$. That means that $J(x) x^{-s-1}$ grows slower than x^{-s}. We avoid any problems near 0 because $J(x) = 0$ for $x < 2$. Thus, for $Re(s) > 1$, we can use the inverse Laplace transform on equation (10.4) and obtain

$$J(x) = \frac{1}{2\pi i} \int_{a-i\infty}^{a+i\infty} \log \zeta(s) \frac{x^s}{s} \, ds \quad a > 1.$$

It turns out, this form of the integral does not work for our purposes. However, if we integrate by parts, we can obtain an integral in the form we need. Let $u(s) = \log \zeta(s)/s$ and $v(s) = x^s/\log x$, we have:

$$\int_{a-i\infty}^{a+i\infty} \frac{\log \zeta(s)}{s} x^s \, ds = \int_{a-i\infty}^{a+i\infty} u(s) v'(s) \, ds = u(s) v(s) \Big|_{a-i\infty}^{a+i\infty} - \int_{a-i\infty}^{a+i\infty} u'(s) v(s) \, ds$$
$$= \frac{\log \zeta(s)}{s} \frac{x^s}{\log x} \Big|_{a-i\infty}^{a+i\infty} - \frac{1}{\log x} \int_{a-i\infty}^{a+i\infty} \frac{d}{ds} \left[\frac{\log \zeta(s)}{s} \right] x^s \, ds, \tag{10.5}$$

where the left expression in equation (10.5) evaluates to 0. This allows us to restate $J(x)$ as:

$$J(x) = -\frac{1}{2\pi i} \cdot \frac{1}{\log x} \int_{a-i\infty}^{a+i\infty} \frac{d}{ds} \left[\frac{\log \zeta(s)}{s} \right] x^s \, ds \quad x > 1, a > 1. \tag{10.6}$$

To see that the integral converges, it is enough to show

$$\lim_{T \to \infty} \frac{\log \zeta(a \pm iT)}{a \pm iT} x^{a \pm iT} = 0.$$

First, $|x^{a \pm iT}| = e^{a \log x}$ is bounded. Next, for the denominator, $\lim_{T \to \infty} |a \pm iT| \to \infty$. Finally,

$$|\log \zeta(a \pm iT)| = \left| \sum_p \left(\sum_n \frac{1}{n} \cdot p^{-n(a \pm iT)} \right) \right| \le \sum_p \left(\sum_n \frac{1}{n} \cdot |p^{-n(a \pm iT)}| \right) = \sum_p \left(\sum_n \frac{1}{n} \cdot p^{-na} \right) = \log \zeta(a),$$

so the numerator is bounded, giving our desired result.

10.2.3 An Equation for $\log \zeta(s)$

We have shown a relationship between $J(x)$ and $\log \zeta(s)$. We now want to develop a more useful equation for $\log \zeta(s)$.

We start with two different definitions of $\xi(s)$, equations (8.1) and (8.4):

$$\xi(s) = \frac{s(s-1)}{2} \pi^{\frac{-s}{2}} \Gamma\left(\frac{s}{2}\right) \zeta(s) = (s-1) \pi^{\frac{-s}{2}} \left[\frac{s}{2} \Gamma\left(\frac{s}{2}\right)\right] \zeta(s)$$

$$= (s-1) \pi^{\frac{-s}{2}} \Gamma\left(\frac{s}{2}+1\right) \zeta(s)$$

$$\xi(s) = \xi(0) \prod_\rho \left(1 - \frac{s}{\rho}\right).$$

In the final definition, ρ has its usual meaning (and ordering) as the roots of $\xi(s)$ (see page xi) and ρ and $1 - \rho$ are paired. Now use $\log \zeta(s) = \log \xi(s) - [\log \xi(s) - \log \zeta(s)]$ to obtain

$$\log \zeta(s) = \log\left(\xi(0) \prod_\rho \left(1 - \frac{s}{\rho}\right)\right) - \left[\log\left((s-1) \pi^{\frac{-s}{2}} \Gamma\left(\frac{s}{2}+1\right) \zeta(s)\right) - \log \zeta(s)\right]$$

$$\log \zeta(s) = \log \xi(0) + \sum_\rho \log\left(1 - \frac{s}{\rho}\right) - \log\left(\Gamma\left(\frac{s}{2}+1\right)\right) + \frac{s}{2} \log \pi - \log(s-1). \qquad (10.7)$$

10.3 The Strategy

Riemann's goal was to develop an *explicit formula* for $\pi(x)$. We have the basic building blocks from above. We have in equation (10.3) a formula for $\pi(x)$:

$$\pi(x) = \sum_{n=1}^\infty \frac{\mu(n)}{n} J(x^{1/n}).$$

We have in equation (10.6) a formula for $J(x)$, but not yet in a form useful for computing $\pi(x)$:

$$J(x) = -\frac{1}{2\pi i} \cdot \frac{1}{\log x} \int_{a-i\infty}^{a+i\infty} \frac{d}{ds}\left[\frac{\log \zeta(s)}{s}\right] x^s \, ds \quad a > 1.$$

And we have in equation (10.7) an alternate expression for $\log \zeta(s)$:

$$\log \zeta(s) = \log \xi(0) + \sum_\rho \log\left(1 - \frac{s}{\rho}\right) - \log\left(\Gamma\left(\frac{s}{2}+1\right)\right) + \frac{s}{2} \log \pi - \log(s-1).$$

The strategy is to plug the equation (10.7) formula for $\log \zeta(s)$ into the equation (10.6) formula for $J(x)$, and then evaluate the $J(x)$ integral term-by-term. We do that in sections 10.4 through 10.8. In all that follows, we assume $s \in \mathbb{C}$, $x > 1$. We also assume that any log function follows the principal branch of the logarithm.

10.4 The Principal Term

We begin by evaluating the $-\log(s-1)$ term of equation (10.7):

$$\frac{1}{2\pi i} \cdot \frac{1}{\log x} \int_{a-i\infty}^{a+i\infty} \frac{d}{ds}\left[\frac{\log(s-1)}{s}\right] x^s \, ds. \qquad (10.8)$$

To assist in our evaluation, we define three helper functions.

Definition 10.1. *Fix $x > 1$, assume $Re(s) > 1$ and let $\beta \in \mathbb{C}\backslash(-\infty, 0]$, $a > 1$ and $a > Re(\beta)$. For small $\epsilon > 0$, let $C+$ be the contour that consists of the real line segment $[0, 1 - \epsilon]$, the semicircle in the upper half-plane from $1 - \epsilon$ to $1 + \epsilon$, and then the real line segment $[1 + \epsilon, x]$. We define*

$$F(\beta) = \frac{1}{2\pi i} \cdot \frac{1}{\log x} \int_{a-i\infty}^{a+i\infty} \frac{d}{ds}\left[\frac{\log\left[(s/\beta) - 1\right]}{s}\right] x^s \, ds,$$

$$G(\beta) = \int_{C+} \frac{t^{\beta-1}}{\log t} \, dt \qquad Re(\beta) > 0,$$

$$H(\beta) = \frac{1}{2\pi i} \cdot \frac{1}{\log x} \int_{a-i\infty}^{a+i\infty} \frac{d}{ds}\left[\frac{\log\left[1 - (s/\beta)\right]}{s}\right] x^s \, ds.$$

In the following lemma, we establish various properties of our helper functions.

Lemma 10.1. *Let $F(\beta)$, $G(\beta)$ and $H(\beta)$ be as defined in Definition 10.1. We have:*

(i) $F(1) = \dfrac{1}{2\pi i} \cdot \dfrac{1}{\log x} \int_{a-i\infty}^{a+i\infty} \dfrac{d}{ds}\left[\dfrac{\log(s-1)}{s}\right] x^s \, ds.$ (i.e., Equation (10.8)).

(ii) $F'(\beta) = x^\beta / \beta.$

(iii) $G'(\beta) = F'(\beta).$

(iv) $\displaystyle\lim_{Im(\beta)\to\infty} G(\beta) = 0$ *for fixed $Re(\beta) > 0$.*

(v)   *For $Im(\beta) > 0$, $H(\beta) - F(\beta) = -i\pi$.*

(vi) $F(\beta) = G(\beta) + i\pi$ *for $Re(\beta) > 0$.*

Proof (i). This result is immediate by direct substitution. □

Proof (ii). First note that the $F(\beta)$ integral is absolutely convergent because the term

$$\left|\frac{d}{ds}\frac{\log\left[(s/\beta) - 1\right]}{s}\right| \leq \frac{\left|\log\left[(s/\beta) - 1\right]\right|}{|s|^2} + \frac{1}{|s(s - \beta)|}$$

is convergent for all β and $|x^s| = e^{a\log x}$. Next, we evaluate

$$\frac{d}{d\beta}\left[\frac{\log\left[(s/\beta) - 1\right]}{s}\right] = \frac{d}{d\beta}\left[\frac{\log(s-\beta) - \log\beta}{s}\right] = \frac{1}{s}\left[\frac{-1}{s-\beta} - \frac{1}{\beta}\right] = \frac{1}{(\beta-s)\beta}.$$

We can now differentiate under the integral sign and then use integration by parts to obtain

$$F'(\beta) = \frac{1}{2\pi i} \cdot \frac{1}{\log x} \int_{a-i\infty}^{a+i\infty} \frac{d}{ds}\left[\frac{1}{(\beta-s)\beta}\right] x^s \, ds$$

$$= -\frac{1}{2\pi i} \int_{a-i\infty}^{a+i\infty} \frac{x^s}{(\beta-s)\beta} \, ds$$

$$= \frac{1}{2\pi i\beta} \int_{a-i\infty}^{a+i\infty} \frac{x^s}{(s-\beta)} \, ds.$$

Applying lemma 12.3, we have our result:

$$= \frac{x^\beta}{\beta}.$$ □

Proof (iii). $\frac{d}{d\beta} t^{\beta-1} = \left(t^{\beta-1} \log t\right)$ so that $G'(\beta) = \int_{C+} t^{\beta-1} dt = \frac{t^\beta}{\beta}\Big|_0^x = \frac{x^\beta}{\beta} = F'(\beta).$ □

Proof (iv). For fixed $Re(\beta) > 0$, and fixed $x > 1$, we must show

$$\lim_{Im(\beta)\to\infty} G(\beta) = \lim_{Im(\beta)\to\infty} \int_{C+} \frac{t^{\beta-1}}{\log t} dt = 0.$$

We would like to make the change of variable $t = e^u$. The problem is that our lower bound is 0 and there is no value of u satisfying $0 = e^u$. Edwards manages this by setting the lower bound to $i\delta - \infty$, with the implied assumption that $\delta \to 0$. Our proof is essentially the same as Edwards', but we will more directly deal with the limit assumption.

For $N \in \mathbb{N}$, let C_N be the contour from $e^{i\delta-N}$ to $e^{i\delta+\log x}$ to $e^{\log x}$, with $\delta = 1/N$. Now define

$$G_N(\beta) = \int_{C_N} \frac{t^{\beta-1}}{\log t} dt \qquad \text{so that} \qquad \lim_{N\to\infty} G_N(\beta) = G(\beta).$$

Let $\beta = A + i\tau$ for fixed $A > 0$, let $\tau \to \infty$, and make the change of variable $t = e^u$:

$$G_N(\beta) = \int_{i\delta-N}^{i\delta+\log x} \frac{e^{\beta u}}{u} du + \int_{i\delta+\log x}^{\log x} \frac{e^{\beta u}}{u} du.$$

Now make the change of variable $u = i\delta + v$ in the first integral, and $u = \log x + iw$ in the second

$$G_N(\beta) = e^{i\delta A} e^{-\delta\tau} \int_{-N}^{\log x} \frac{e^{Av}}{i\delta + v} e^{i\tau v} dv - ix^\beta \int_0^\delta \frac{e^{-\tau w} e^{Aiw}}{\log x + iw} dw,$$

We now evaluate the two integrals separately. For the first integral, we use the ML inequality

$$\leq e^{-\delta\tau} \left|e^{i\delta A}\right| \left|\int_{-N}^{\log x} \frac{e^{Av}}{i\delta + v} e^{i\tau v} dv\right| \leq e^{-\delta\tau} (N + \log x) \cdot \max_{-N \leq v \leq \log x} \left|\frac{e^{Av}}{i\delta + v}\right|$$

$$\leq e^{-\delta\tau} (N + \log x) \frac{e^{A\log x}}{\delta} = e^{-\tau/N} (N + \log x) \left(Ne^{A\log x}\right).$$

With all other values in our last equation fixed, the first integral goes to 0 as $\tau \to \infty$.

We now evaluate the second integral. With $\left|ix^\beta\right| = x^A$, that term is fixed and can be disregarded. Note that $e^{-\tau w} \to 0$ except at $w = 0$ (a point of measure zero). Let $M \in \mathbb{N}$ with $M > N$. Then

$$\int_0^\delta \frac{e^{-\tau w} e^{Aiw}}{\log x + iw} dw = \lim_{M\to\infty} \int_{1/M}^{1/N} \frac{e^{-\tau w} e^{Aiw}}{\log x + iw} dw.$$

For any given M we can use the ML Inequality and have

$$\int_{1/M}^{1/N} \frac{e^{-\tau w} e^{Aiw}}{\log x + iw} dw \leq (1/N - 1/M) \cdot \max_{1/M \leq w \leq 1/N} \left|\frac{e^{-\tau w} e^{Aiw}}{\log x + iw}\right| \leq (1/N - 1/M) \frac{e^{-\tau/M}}{\log x}.$$

With all other values in our last equation fixed, the second integral goes to 0 as $\tau \to \infty$. Therefore, the limit of $G(\beta)$ as $\tau \to \infty$ is 0. □

Proof (v). In the upper half plane, where $Im(\beta) > 0$, we have

$$H(\beta) - F(\beta) = \frac{1}{2\pi i} \cdot \frac{1}{\log x} \int_{a-i\infty}^{a+i\infty} \frac{d}{ds}\left(\frac{\log(s-\beta) - \log(-\beta) - [\log(s-\beta) - \log\beta]}{s} \right) x^s\, ds,$$

$$= \frac{1}{2\pi i} \cdot \frac{1}{\log x} \int_{a-i\infty}^{a+i\infty} \frac{d}{ds}\left[\frac{\log\beta - \log(-\beta)}{s} \right] x^s\, ds$$

$$= \frac{1}{2\pi i} \cdot \frac{1}{\log x} \int_{a-i\infty}^{a+i\infty} \frac{d}{ds}\left[\frac{i\pi}{s} \right] x^s\, ds \qquad (\text{using } i\operatorname{Arg}(\beta) - i\operatorname{Arg}(-\beta) = i\pi)$$

$$= -\frac{1}{2\pi i} \int_{a-i\infty}^{a+i\infty} \left[\frac{i\pi}{s} \right] x^s\, ds$$

$$= -i\pi.$$

The last result follows from lemma 12.3, by setting $\beta = 0$. $\qquad\square$

Proof (vi). Let $\beta = A + i\tau$. Fix $A > 0$, let $a > A$, and evaluate $F(\beta)$ as $\tau \to \infty$.

In lemma 10.1(v), we showed that $F(\beta) = H(\beta) + i\pi$ in the upper half plane. Next, we find the limit of $H(\beta)$ as $\tau \to \infty$. We have

$$\frac{d}{ds}\left[\frac{\log[1-(s/\beta)]}{s} \right] = \frac{-s/(\beta - s)}{s^2} - \frac{\log[1-(s/\beta)]}{s^2} = -\frac{\log[1-(s/\beta)]}{s^2} + \frac{1}{s(s-\beta)}$$

$$= -\frac{\log[1-(s/\beta)]}{s^2} + \frac{1}{\beta(s-\beta)} - \frac{1}{\beta s},$$

which may be substituted in the integral defining $H(\beta)$. So we have

$$H(\beta) = \frac{1}{2\pi i} \cdot \frac{1}{\log x} \int_{a-i\infty}^{a+i\infty} \left[-\frac{\log[1-(s/\beta)]}{s^2} + \frac{1}{\beta(s-\beta)} - \frac{1}{\beta s} \right] x^s\, ds.$$

Considering only the first substituted term, we have

$$\frac{1}{2\pi i} \cdot \frac{1}{\log x} \int_{a-i\infty}^{a+i\infty} \left[-\frac{\log[1-(s/\beta)]}{s^2} \right] x^s\, ds.$$

Fix $a + iT$ along the line of integration. We first show that, as $\tau \to \infty$, $s/b \to 0$.

$$\lim_{\tau\to\infty}\left[\frac{s}{b} = \frac{(a+iT)}{(A+i\tau)} \cdot \frac{(A-i\tau)}{(A-i\tau)} = \frac{aA + iAT - ia\tau + T\tau}{A^2 + \tau^2} \approx \frac{ia\tau + T\tau}{\tau^2} = \frac{ia+T}{\tau} \right] \to 0.$$

Thus, as $\tau \to \infty$, $[1-(s/\beta)] \to 1$ and $\log[1-(s/\beta)] \to 0$. The denominator grows like $|s|^2$ for large τ, and $|x^s| = e^{a\log x}$. We are therefore in a position to use the Lebesgue bounded convergence theorem, so that the limit of the integral equals the integral of the limit. With a limit of 0, our integral is also 0.

Turning to the second and third substituted terms, we have

$$\frac{1}{2\pi i} \cdot \frac{1}{\log x} \int_{a-i\infty}^{a+i\infty} \left[\frac{1}{\beta(s-\beta)} - \frac{1}{\beta s} \right] x^s\, ds = \frac{1}{\log x}\left[\frac{x^\beta}{\beta} - \frac{1}{\beta} \right] = \frac{x^\beta - 1}{\beta \log x}$$

using lemma 12.3. Because $Re(\beta) < a$ and x is fixed, the numerator is bounded and $|\beta| \to \infty$, so these terms go to 0, and the function $H(\beta)$ goes to 0 as $\tau \to \infty$. But that means $F(\beta)$ goes to $i\pi$ as $\tau \to \infty$. Therefore, $F(\beta) = G(\beta) + i\pi$ for $Re(\beta) > 0$. $\qquad\square$

With our three helper functions and their properties, we have enough to evaluate equation (10.8) (the "Principal Term" of our integral for $J(x)$). From Lemma 10.1(i), $F(1)$ is equal to equation (10.8). From Lemma 10.1(iii), $F'(\beta) = G'(\beta)$, which means that $F(\beta)$ and $G(\beta)$ must differ by a constant. In fact, we determined that constant in Lemma 10.1(vi) and have

$$F(\beta) = G(\beta) + i\pi \text{ for } Re(\beta) > 0.$$

Putting it all together, we can compute $F(1)$, our Principal Term:

$$\begin{aligned} F(1) &= G(1) + i\pi \\ &= \int_{C+} \frac{dt}{\log t} + i\pi \\ &= \int_0^{1-\epsilon} \frac{dt}{\log t} + \int_{1-\epsilon}^{1+\epsilon} \frac{dt}{\log t} + \int_{1+\epsilon}^x \frac{dt}{\log t} + i\pi. \end{aligned} \quad (10.9)$$

Taking the limit as $\epsilon \to 0$, we see that the second term approaches along a pole of residue 1, but the contour is taken with the negative orientation, resulting a value of $-i\pi$ from the residue theorem. This implies that the second integral and the last term cancel, and we are left with

$$\boxed{F(1)} = \lim_{\epsilon \to 0} \int_0^{1-\epsilon} \frac{dt}{\log t} + \int_{1+\epsilon}^x \frac{dt}{\log t} = \boxed{li(x)}.$$

10.5   The Oscillating Term

We next consider the following term of equation (10.7):

$$\sum_\rho \log\left(1 - \frac{s}{\rho}\right).$$

As a term in the $J(x)$ integral, this is

$$-\frac{1}{2\pi i} \cdot \frac{1}{\log x} \int_{a-i\infty}^{a+i\infty} \frac{d}{ds}\left[\frac{\sum_\rho \log\left(1 - \frac{s}{\rho}\right)}{s}\right] x^s \, ds. \quad (10.10)$$

This term causes Riemann the most difficulties. The obvious (and perhaps only) approach is to interchange the sum and the integral/differentiation. But it is far from obvious that the necessary conditions are met to justify the interchange. Riemann forges ahead, but in his private papers he acknowledges that justifying the interchange was an unresolved issue. We follow Riemann.

$$-\sum_\rho \frac{1}{2\pi i} \cdot \frac{1}{\log x} \int_{a-i\infty}^{a+i\infty} \frac{d}{ds}\left[\frac{\log\left(1 - \frac{s}{\rho}\right)}{s}\right] x^s \, ds = -\sum_\rho H(\rho),$$

where $H(\rho)$ is as defined in Definition 10.1. If you combine the results of Lemma 10.1(v) & (vi), we have that $H(\rho) = G(\rho)$ in the first quadrant ($Re(\rho) > 0, Im(\rho) > 0$). Now modify the contour $C+$ used in the $G(\rho)$ integral to a new contour $C-$ which is identical to $C+$ except the semicircle between $1 - \epsilon$ and $1 + \epsilon$ is in the lower half plane ($Im(\rho) < 0$). It easily follows that $H(\rho) = G(\rho)$ in the fourth quadrant ($Re(\rho) > 0, Im(\rho) < 0$).

Our sum over ρ has its usual meaning (and ordering) as the roots of $\xi(s)$ (see page xi). In this case, we also pair the terms ρ and $1 - \rho$, allowing a sum over $Im(\rho) > 0$. We have

$$-\sum_\rho H(\rho) = - \sum_{Im(\rho)>0} \left(\int_{C+} \frac{t^{\rho-1}}{\log t} dt + \int_{C-} \frac{t^{-\rho}}{\log t} dt \right). \tag{10.11}$$

For the first integral, temporarily assume $\{\beta \in \mathbb{R} : \beta > 0\}$ and make the change of variable $u = t^\beta$, so that $\log t = \log u / \beta$, $dt/t = du/u\beta$, so that, for the reasons explained after equation (10.9):

$$\int_{C+} \frac{t^{\beta-1}}{\log t} dt = \int_0^{x^\beta} \frac{du}{\log u} = li(x^\beta) - i\pi,$$

where the path of the right integral stays above the real line near $u = 1$. Since both integrals converge for $\{\beta \in \mathbb{C} : Re(\beta) > 0\}$, the right integral gives an analytic continuation of $li(x^\beta)$ for all $Re(\beta) > 0$. Similarly, if we pair the path $C-$ with a path that stays *below* the real line near $u = 1$, we have

$$\int_{C-} \frac{t^{\beta-1}}{\log t} dt = \int_0^{x^\beta} \frac{du}{\log u} = li(x^\beta) + i\pi.$$

Combining our results, we have for this term of $J(x)$

$$\boxed{- \sum_{Im(\rho)>0} \left[li(x^\rho) + li(x^{1-\rho}) \right]}. \tag{10.12}$$

As mentioned above, we have nowhere justified the interchange (termwise evaluation of the integral). It was not until 1895 that von Mangoldt [26] proved (indirectly) that termwise evaluation led to the correct result. And it was not until 1908 that Landau [23] provided a direct proof.

10.6   The Constant Term

We next consider the $\log \xi(0)$ term of equation (10.7) and insert that term in the integral:

$$-\frac{1}{2\pi i} \cdot \frac{1}{\log x} \int_{a-i\infty}^{a+i\infty} \frac{d}{ds} \left[\frac{\log \xi(0)}{s} \right] x^s ds. \tag{10.13}$$

Using integration by parts and lemma 12.3 (with $\beta = 0$), we have

$$= -\frac{1}{2\pi i} \cdot \frac{1}{\log x} \left[\frac{\log \xi(s)}{s} x^s \Big|_{a-i\infty}^{a+i\infty} - \int_{a-i\infty}^{a+i\infty} \left(\frac{\log \xi(0)}{s} \right) \left(\frac{d}{ds} x^s \right) ds \right]$$

$$= -\frac{1}{2\pi i} \cdot \frac{1}{\log x} \left[0 - \int_{a-i\infty}^{a+i\infty} \left(\frac{\log \xi(0)}{s} \right) (x^s \log x) \, ds \right]$$

$$= \log \xi(0) \left[\frac{1}{2\pi i} \int_{a-i\infty}^{a+i\infty} \frac{x^s}{(s-0)} ds \right] = \log \xi(0) \cdot x^0$$

$$= \log \xi(0).$$

Now using

$$\xi(s) = (s-1) \pi^{\frac{-s}{2}} \Gamma\left(\frac{s}{2} + 1\right) \zeta(s)$$

$$\xi(0) = (0-1) \pi^0 \Gamma(1) \zeta(0) = \zeta(0) = 1/2,$$

we have $\log \xi(0) = \boxed{-\log 2}$.

10.7   The Gamma Term

We next consider the $\log\left(\Gamma\left(\frac{s}{2}+1\right)\right)$ term of equation (10.7) and insert that term in the integral:

$$-\frac{1}{2\pi i}\cdot\frac{1}{\log x}\int_{a-i\infty}^{a+i\infty}\frac{d}{ds}\left[\frac{\log\left(\Gamma\left(\frac{s}{2}+1\right)\right)}{s}\right]x^s\,ds. \tag{10.14}$$

Using the formula

$$\Gamma(s+1)=\prod_{n=1}^{\infty}\left(1+\frac{s}{n}\right)^{-1}\left(1+\frac{1}{n}\right)^{s},$$

we have

$$\log\left(\Gamma\left(\frac{s}{2}+1\right)\right)=\sum_{n=1}^{\infty}\left[-\log\left(1+\frac{s}{2n}\right)+\frac{s}{2}\log\left(1+\frac{1}{n}\right)\right].$$

Putting it together, this term of $J(x)$ is

$$=-\frac{1}{2\pi i}\cdot\frac{1}{\log x}\int_{a-i\infty}^{a+i\infty}\frac{d}{ds}\left[\frac{\sum_{n=1}^{\infty}\left[-\log\left(1+\frac{s}{2n}\right)\right]}{s}+\frac{\sum_{n=1}^{\infty}\left[\frac{s}{2}\log\left(1+\frac{1}{n}\right)\right]}{s}\right]x^s\,ds$$

$$=-\frac{1}{2\pi i}\cdot\frac{1}{\log x}\int_{a-i\infty}^{a+i\infty}\frac{d}{ds}\left[\frac{\sum_{n=1}^{\infty}\left[-\log\left(1+\frac{s}{2n}\right)\right]}{s}+\sum_{n=1}^{\infty}\left[\frac{1}{2}\log\left(1+\frac{1}{n}\right)\right]\right]x^s\,ds.$$

But the second sum is a constant, and the derivative of a constant is 0, so we simplify to

$$=-\frac{1}{2\pi i}\cdot\frac{1}{\log x}\int_{a-i\infty}^{a+i\infty}\frac{d}{ds}\left[\frac{\sum_{n=1}^{\infty}\left[-\log\left(1+\frac{s}{2n}\right)\right]}{s}\right]x^s\,ds. \tag{10.15}$$

We will justify the interchange of sum and integral below. For now, we restate the above equation as

$$=-\sum_{n=1}^{\infty}H(-2n), \tag{10.16}$$

where $H(\beta)$ is as defined in Definition 10.1. There, we only evaluated $H(\beta)$ for $Re(\beta)>0$. Here we study $H(\beta)$ for $Re(\beta)<0$. Define

$$E(\beta)=-\int_{x}^{\infty}\frac{t^{\beta-1}}{\log t}\,dt,$$

so that

$$E'(\beta)=-\int_{x}^{\infty}t^{\beta-1}\,dt=\frac{x^{\beta}}{\beta}=H'(\beta).$$

We see that $E(\beta)$ and $H(\beta)$ differ by a constant. That constant is 0 because both $E(\beta)$ and $H(\beta)\to 0$ as $\beta\to\infty$. Thus, $E(\beta)=H(\beta)$, and our term becomes

$$-\sum_{n=1}^{\infty}H(-2n)=\sum_{n=1}^{\infty}\int_{x}^{\infty}\frac{t^{-2n-1}}{\log t}\,dt=\sum_{n=1}^{\infty}\int_{x}^{\infty}\left(\frac{1}{t\log t}\right)t^{-2n}\,dt=\int_{x}^{\infty}\left(\frac{1}{t\log t}\right)\left(\sum_{n=1}^{\infty}t^{-2n}\right)dt.$$

Since $t > 1$, we can use the formula for a geometric series and have

$$\sum_{n=1}^{\infty} t^{-2n} = \left(\sum_{n=0}^{\infty} t^{-2n} \right) - 1 = \left(\sum_{n=0}^{\infty} \left(t^{-2} \right)^n \right) - 1 = \frac{1}{1 - t^{-2}} - 1 = \frac{1}{t^2 - 1},$$

so that our $J(x)$ term becomes

$$= \int_x^{\infty} \left(\frac{1}{t \log t} \right) \left(\frac{1}{t^2 - 1} \right) dt = \boxed{\int_x^{\infty} \frac{dt}{t(t^2 - 1) \log t}}.$$

It remains to justify the interchange of sum, derivative and integral between equations (10.15) and (10.16): $\int d/ds (\sum f(s)) \, ds \to \int \sum (d/ds f(s)) \, ds \to \sum \int (d/ds f(s)) \, ds$.

To justify termwise differentiation, we must show that

$$\sum_{n=1}^{\infty} \left[\frac{d}{ds} \frac{-\log (1 + s/2n)}{s} \right]$$

converges uniformly. For sufficiently large n, we can use the Taylor Series expansion

$$\log(1 + x) = x - \frac{1}{2}x^2 + \frac{1}{3}x^3 - \ldots \qquad |x| < 1.$$

We have $x = s/2n$ so that

$$\frac{d}{ds} \frac{-\log (1 + s/2n)}{s} = \frac{d}{ds} - \left(\frac{1}{s} \left[\frac{s}{2n} - \frac{1}{2} \frac{s^2}{4n^2} + \frac{1}{3} \frac{s^3}{8n^3} - \ldots \right] \right)$$

$$= \frac{d}{ds} - \left[\frac{1}{2n} - \frac{1}{2} \frac{s}{4n^2} + \frac{1}{3} \frac{s^2}{8n^3} - \ldots \right]$$

$$= \frac{1}{2} \frac{1}{4n^2} - \frac{2}{3} \frac{s}{8n^3} + \frac{3}{4} \frac{s^2}{16n^4} - \ldots,$$

which converges uniformly because the highest order term of n is n^{-2}.

It is a bit more delicate to justify termwise integration. Basically, you integrate over a finite interval $[a - iT, a + iT]$, show (with a change of variable and integration by parts) that for all T there is decay on the order of n^{-2}, and then pass $T \to \infty$ to the limit. This is nicely done by Edwards [13, 32-33].

10.8   The Vanishing Term

Finally, we consider the $\frac{s}{2} \log \pi$ term of equation (10.7) and insert that term in the integral:

$$-\frac{1}{2\pi i} \cdot \frac{1}{\log x} \int_{a-i\infty}^{a+i\infty} \frac{d}{ds} \left[\frac{\frac{s}{2} \log \pi}{s} \right] x^s \, ds = \boxed{0}.$$

When divided by s, the term becomes a constant, with a derivative of 0. Thus, the entire term (and integral) are 0.

10.9　Putting it All Together

We have shown that for $x > 1$

$$J(x) = li(x) - \sum_{Im(\rho)>0} \left[li(x^\rho) + li(x^{1-\rho}) \right] + \int_x^\infty \frac{dt}{t(t^2-1)\log t} . - \log 2, \qquad (10.17)$$

Our above equation, along with the equation (10.2) definition of $\mu(n)$, allows us to restate equation (10.3) as our final result:

$$\pi(x) = \sum_{n=1}^\infty \frac{\mu(n)}{n} J(x^{1/n}). \qquad (10.18)$$

From equation (10.1), it is clear that $J(x) = 0$ for $x^{1/n} < 2$. Thus, the series in equation (10.18) is finite for any given x. More specifically,

$$\pi(x) = \sum_{n=1}^N \frac{\mu(n)}{n} J(x^{1/n}) \qquad \text{for } x < 2^N.$$

This was Riemann's main goal in the paper – an analytic formula for $\pi(x)$.

Now consider the four terms defining $J(x)$ in equation (10.17). The third and fourth terms do not increase with x. The second term is very difficult to evaluate – it *could* increase with x, but when computed it appears to oscillate in sign (Riemann calls it "periodic"). Of course, the first term increases consistently with x. Disregarding the last three terms, we have

$$\pi(x) \sim li(x) - \frac{1}{2} li(x^{1/2}) - \frac{1}{3} li(x^{1/3}) - \frac{1}{5} li(x^{1/5}) + \frac{1}{6} li(x^{1/6}) - \frac{1}{7} li(x^{1/7}) + ...,$$

which has shown to be a very good approximation of $\pi(x)$. Of course, that means we have an exact analytic formula for the approximation error for any given $x < 2^N$:

$$\pi(x) - \sum_{n=1}^N \frac{\mu(n)}{n} li(x^{1/n}) = \sum_{n=1}^N \sum_{Im(\rho)>0} \left[li(x^{\rho/n}) + li(x^{(1-\rho)/n}) \right] + \text{ lesser terms.}$$

As mentioned, the *empirical* evidence shows that the error term is relatively small. Looking at the *oscillating term*, it is far from clear why that should be. Riemann's only comment is to say that: "[no] law governing this behavior [has] been observed".

Entire Functions

An assumption by Riemann (central to Riemann's Paper) is that $\xi(s)$ can be expressed in terms of an infinite product over its zeros, a fact left unproven in his paper. In fact, the theory supporting his conclusion was not fully developed until Jacques Hadamard proved *Hadamard's Factorization Theorem* in 1893 [17].

We will develop here[30] enough of the theory of entire functions to allow proof of *Hadamard's Factorization Theorem*. The obvious starting point is a study of infinite products in the complex plane. Another focal point will be the critical contributions made by Weierstrass in the 1870's.

11.1 Infinite Products

Let $\{a_k\}$ be a sequence of complex numbers and denote $P_n = \prod_{k=1}^n (a_k)$ as the nth *partial product* of $\{a_k\}$. We say the infinite product $\prod_{k=1}^\infty (a_k)$ *converges* if $\lim_{n \to \infty} P_n = P$ for some non-zero complex number P.

The above definition excludes the case where one or more of the $\{a_k\} = 0$, in which case the product equals 0 without regard to the values of the other terms. However, it is sometimes still useful to evaluate such products by excluding the $\{a_k\} = 0$ terms, *but only if there are a finite number of such terms*. We assume here that all terms of the product are non-zero unless specifically stated otherwise.

Now consider the sequence $\{P_n\}$ where $P_n \to P \neq 0$, and assume it is convergent. Then we must have $P_n/P_{n-1} \to P/P = 1$. By definition, $P_n/P_{n-1} = a_n$. Therefore, for $\prod_{k=1}^\infty (a_k)$ to converge, a necessary (but not sufficient) condition is that $a_k \to 1$ as $k \to \infty$. For this reason, it is customary to write the infinite product as $\prod_{k=1}^\infty (1 + a_k)$, with convergence possible only if $a_k \to 0$ as $k \to \infty$.

11.1.1 Convergence – Three Simple Lemmas

We start with a simple real-variable case.

Lemma 11.1. *If $\{a_n\} \in \mathbb{R}_{\geq 0}$, then $\sum_{n=1}^\infty a_n$ converges if and only if $\prod_{n=1}^\infty (1 + a_n)$ converges.*

Proof. We first show that for $x \in \mathbb{R}_{\geq 0}$, we have $1 + x \leq e^x$:

$$1 + x = 1 + \left(n \cdot \frac{x}{n}\right) \leq \left(1 + \frac{x}{n}\right)^n \leq \lim_{n \to \infty} \left(1 + \frac{x}{n}\right)^n = e^x.$$

Thus, for all n:

$$a_1 + a_2 + \ldots a_n \leq (1 + a_1) \cdot (1 + a_2) \cdots (1 + a_n) \leq e^{a_1} \cdot e^{a_2} \cdots e^{a_n} = e^{a_1 + a_2 + \ldots a_n}. \qquad \square$$

We use the above result to prove two simple complex-variable cases.

Lemma 11.2. *If $\{a_n\} \in \mathbb{C}$, then: (i) the sum $\sum_{n=1}^{\infty} |a_n|$ converges if and only if the product $\prod_{n=1}^{\infty}(1 + |a_n|)$ converges, and (ii) if the sum $\sum_{n=1}^{\infty} |a_n|$ converges, then the product $\prod_{n=1}^{\infty} |1 + a_n|$ converges.*

Proof. Because $|a_n|$ is real-valued, (i) is immediate from Lemma 11.1. For (ii), we again use Lemma 11.1 and note by the triangle inequality that $|1 + a_n| \leq |1| + |a_n| = (1 + |a_n|)$. $\square$

Lemma 11.3. *For $\{z_k\} \in \mathbb{C}$, we have $\left|\prod_{k=1}^{n}(1 + z_k) - 1\right| \leq \prod_{k=1}^{n}(1 + |z_k|) - 1$.*

Proof. Our proof will be by induction.
Base Case: $n = 1$. The inequality is easily seen to be true for $n = 1$.
Assumed Case: $n = m$. We assume the inequality is true for $n = m$. Let $A = \prod_{k=1}^{m}(1 + z_k)$ and let $B = \prod_{k=1}^{m}(1 + |z_k|)$. Therefore $|A - 1| \leq B - 1$. By the triangle inequality we have $|A| - |1| \leq |A - 1|$ so that $|A| \leq |A - 1| + |1| \leq B - 1 + |1| \Longrightarrow |A| \leq B$.
Induction Step: $n = m + 1$. For $z \in \mathbb{C}$, we need to prove $|A(1 + z) - 1| \leq B(1 + |z|) - 1$. Our approach will be to assume the inequality is false and look for a contradiction. We therefore assume

$$B(1 + |z|) - 1 < |A(1 + z) - 1|$$

so that

$$
\begin{aligned}
(B - 1) + B|z| &< |A - 1 + Az| \\
&< |A - 1| + |Az| &&\text{by the triangle inequality} \\
&< (B - 1) + |Az| &&|A - 1| \leq B - 1 \text{ is assumed in our } n = m \text{ case} \\
B|z| &< |Az|.
\end{aligned}
$$

And we have reached our contradiction because we were to assume $|A| \leq B$. $\square$

11.1.2 Convergence – The Cauchy Criterion

Lemma 11.4. *Let $\{a_n\} \in \mathbb{C}$ with $a_n \neq -1$. The infinite product $\prod_{n=1}^{\infty}(1 + a_n)$ converges if and only if for any $0 < \epsilon < 1$ there is an $N \in \mathbb{N}$ such that*

$$|(1 + a_{n+1})(1 + a_{n+2}) \cdots (1 + a_m) - 1| < \epsilon$$

for all $m > n \geq N$.

Proof. Let $p_n = \prod_{k=1}^{n}(1 + a_k)$ and let ϵ be given.
Convergence $\Longrightarrow$ Cauchy. Assume the infinite product converges, and let $P = \lim_{n \to \infty} p_n$. Because $\{p_n\}$ is a Cauchy sequence, for our given ϵ there is an N such that

$$|p_n| > \frac{|P|}{2} \quad \text{and} \quad |p_n - p_m| < \epsilon \frac{|P|}{2}$$

for all $m > n \geq N$. Our Cauchy condition is satisfied because for all $m > n \geq N$:

$$
\begin{aligned}
|(1 + a_{n+1})(1 + a_{n+2}) \cdots (1 + a_m) - 1| &= \left|\frac{p_m}{p_n} - 1\right| \\
&= |p_n| \left|\frac{p_m}{p_n} - 1\right| \frac{1}{|p_n|} = |p_m - p_n| \frac{1}{|p_n|} \\
&< \epsilon \frac{|P|}{2} \cdot \frac{2}{|P|} = \epsilon.
\end{aligned}
$$

Cauchy $\Rightarrow$ Convergence. Here we assume that for our given ϵ there is an N such that for all $m > n \geq N$:

$$|(1 + a_{n+1})(1 + a_{n+2}) \cdots (1 + a_m) - 1| = \left| \frac{p_m}{p_n} - 1 \right| < \epsilon.$$

Fix N and define $b_k = p_{N+k}/p_N$ for all $k \in \mathbb{N}$, so that

$$1 - \epsilon < |b_k| < 1 + \epsilon < 2.$$

Therefore, for $m > n \geq 0$

$$\frac{b_m}{b_n} = \frac{p_{N+m}}{p_{N+n}} \quad \text{so that} \quad \left| \frac{b_m}{b_n} - 1 \right| < \epsilon \quad \text{and therefore} \quad |b_m - b_n| < \epsilon |b_n| < 2\epsilon.$$

That means $\{b_n\}$ is a Cauchy sequence. But $b_m/b_n = p_{N+m}/p_{N+n}$, so $\{p_n\}$ is also a Cauchy sequence. Hence, $\{p_n\}$ converges, as required. $\square$

11.1.3 Absolute Convergence

We say that the infinite product $\prod_{n=1}^{\infty}(1 + a_n)$ *converges absolutely* if $\prod_{n=1}^{\infty}(1 + |a_n|)$ converges. Similarly, the infinite series $\sum_{n=1}^{\infty} a_n$ *converges absolutely* if $\sum_{n=1}^{\infty} |a_n|$ converges. In fact, by lemma 11.2, the infinite product $\prod_{n=1}^{\infty}(1 + a_n)$ converges absolutely if and only if the infinite series $\sum_{n=1}^{\infty}(a_n)$ converges absolutely. We explore here some features of absolutely convergent series and products.

Lemma 11.5. *If $\{a_n\} \in \mathbb{C}$ and $\sum_{n=1}^{\infty} |a_n|$ converges, then the product $\prod_{n=1}^{\infty}(1 + a_n)$ converges. Under our usual assumption that all $a_n \neq -1$, the product converges to some $P \neq 0 \in \mathbb{C}$.*

Proof. From lemma 11.2, $\prod_{n=1}^{\infty}(1 + |a_n|)$ converges. Using the inequality of lemma 11.3, convergence of $\prod_{n=1}^{\infty}(1 + a_n)$ is immediate by application of lemma 11.4.

To see that $P \neq 0$, we again apply lemma 11.4 and select an N such that

$$|(1 + a_{n+1})(1 + a_{n+2}) \cdots (1 + a_m) - 1| < 1/2$$

for all $m > n \geq N$. Now use the triangle inequality $|1| - |a| \leq |1 - a| = |a - 1|$, so that

$$|(1 + a_{n+1})(1 + a_{n+2}) \cdots (1 + a_m)| > 1/2$$

for all $m > n \geq N$. By assumption, the product of the first N terms is non-zero. The absolute value of the product of the remaining terms is greater than $1/2$. Thus, the full infinite product must converge to some $P \neq 0$, as required. $\square$

We next show that the commutative laws apply to absolutely convergent infinite series and products.

Lemma 11.6. *Assume that $\prod_{n=1}^{\infty}(1 + a_n)$ converges absolutely.*

Define $\{k_n\}$ to be a sequence in which every positive integer appears exactly once, but in no particular pre-assumed order. So, given an infinite sequence $\{a_n\}$ ($n = 1, 2, 3...$), we can rearrange that sequence into the order $\{a_{k_n}\}$. We have, for all such rearrangements:

(i) $\sum_{n=1}^{\infty} a_{k_n}$ *and* $\prod_{n=1}^{\infty}(1 + a_{k_n})$ *converge.*

(ii) $\sum_{n=1}^{\infty} a_n = \sum_{n=1}^{\infty} a_{k_n}.$

(iii) $\prod_{n=1}^{\infty}(1 + a_n) = \prod_{n=1}^{\infty}(1 + a_{k_n}).$

Proof Definitions. For our given rearrangement sequence $\{k_n\}$, and for any given $N \in \mathbb{N}$:

· define $[N]$ to be the lowest positive whole number such that $\{1, 2, ..., N\} \subseteq \{k_1, k_2, ..., k_{[N]}\}$.

· define $[N']$ to be the lowest positive whole number such that $\{k_1, k_2, ..., k_N\} \subseteq \{1, 2, ..., [N']\}$.

Proof (i). From lemma 11.2, $\sum_{n=1}^{\infty} |a_n|$ converges to some $A \in \mathbb{R}$. Let

$$S_j = \sum_{n=1}^{j} |a_n| \quad \text{and} \quad SR_j = \sum_{n=1}^{j} |a_{k_n}|.$$

Both S_j and SR_j are monotone increasing. Clearly, then, all $S_j \leq A$. By definition, for all $N \in \mathbb{N}$ all terms of SR_N are included in $S_{[N']}$, so that $SR_N \leq S_{[N']} \leq A$. Because SR_j is monotone increasing and bounded above, $\sum_{n=1}^{\infty} |a_{k_n}|$ converges. We can then use lemma 11.5 and lemma 11.1 to conclude that both $\sum_{n=1}^{\infty} a_{k_n}$ and $\prod_{n=1}^{\infty} (1 + a_{k_n})$ converge. $\square$

Proof (ii). We have from lemma 11.5 and lemma 11.1 that $\sum_{n=1}^{\infty} a_n$ converges. Let

$$S_j = \sum_{n=1}^{j} a_n \quad \text{and} \quad SR_j = \sum_{n=1}^{j} a_{k_n}.$$

Fix $\epsilon > 0$. Because S_j is a Cauchy sequence, there is an N such that $m > n \geq N \implies \sum_{i=n}^{m} |a_i| < \epsilon$. Now choose $n > [N]$. Then

$$|S_n - SR_n| = \left| \left(\sum_{j=1}^{N} a_j + \sum_{j=N+1}^{n} a_j \right) - \left(\sum_{j=1, k_j \leq N}^{n} a_{k_j} + \sum_{j=1, k_j > N}^{n} a_{k_j} \right) \right|$$

$$= \left| \left(\sum_{j=1}^{N} a_j + \sum_{j=N+1}^{n} a_j \right) - \left(\sum_{j=1}^{N} a_j + \sum_{j=1, k_j > N}^{n} a_{k_j} \right) \right| < 2\epsilon.$$

Both sums include all terms where the index a_j is $\leq N$. Those terms cancel, leaving only the tails, which can be made arbitrarily small. Thus, both series converge to the same sum. $\square$

Proof (iii). We have from lemma 11.5 that $P = \prod_{n=1}^{\infty} (1 + a_n)$ converges. Let

$$P_j = \prod_{n=1}^{j} (1 + a_n) \quad \text{and} \quad Q_j = \prod_{n=1}^{j} (1 + a_{k_n}).$$

From lemma 11.4, there is an N_0 such that $m > n \geq N_0 \implies \left| \left(\prod_{j=n+1}^{m} (1 + |a_j|) \right) - 1 \right| < \epsilon$, and similarly [see (ii) above] an N_1 such that $m > n \geq N_1 \implies \sum_{i=n}^{m} |a_i| < \epsilon$. Let $N = \max(N_0, N_1)$.

Now choose $n > [N]$. Define $\{b_j\}$ as all members of $\{k_j\}$ where $k_j \leq n$. Define $\{c_j\}$ as all members of $\{k_j\}$ where $k_j \leq N$ Finally, define $\{d_j\}$ as all members of $\{k_j\}$ where $k_j > N$ and $j \leq n$. Note that $\{b_j\} = \{c_j\} + \{d_j\}$.

We have $Q_n = \prod_{j=1}^{n} (1 + a_{k_j}) = \prod_{\{b_j\}} (1 + a_{b_j})$. Next, $P_N = \prod_{j=1}^{N} (1 + a_j) = \prod_{\{c_j\}} (1 + a_{c_j})$. Finally, $R_n = \prod_{\{d_j\}} (1 + a_{d_j})$. Then $Q_n = P_N \cdot R_n$, and $|R_n - 1| < \epsilon$.

CHAPTER 11. ENTIRE FUNCTIONS

Applying lemma 11.3 to R_n, we have $|R_n - 1| \leq \left| \left(\prod_{d_j} (1 + |d_j|) \right) - 1 \right| < \epsilon$. Therefore

$$
\begin{aligned}
|Q_n - P| \leq |Q_n - P_N| + |P_N - P| &= |P_N \cdot R_n - P_N| + |P_N - P| \\
&= |P_N| |R_n - 1| + |P_N - P| \\
&< |P_N| |\epsilon| + |P_N - P|
\end{aligned}
$$

By choosing large enough N and small enough ϵ, we can make the right side as small as we like. Thus, $Q_n \to P$, as required. $\qquad \square$

11.1.4 Convergence and the Logarithm

A common technique in complex analysis is to convert a product into a sum by taking the logarithm. That makes the following lemma quite useful.

Lemma 11.7. *Let $\{a_n\} \in \mathbb{C}$ with $a_n \neq -1$, and assume that the terms of $\log(1 + a_n)$ are from the principal branch of the logarithm. Then the series $\sum_{n=1}^{\infty} \log(1 + a_n)$ converges if and only if the product $\prod_{n=1}^{\infty} (1 + a_n)$ converges.*

Proof.[31] Assume the series converges. Set $S_n = \sum_{k=1}^{n} \log(1 + a_k)$ and $P_n = \prod_{k=1}^{n} (1 + a_k)$. We have $P_n = e^{S_n}$ and have assumed $S_n \to S$ for some S. It follows that P_n tends to the limit $P = e^S \neq 0$. So, the convergence of the series assures the convergence of the product.

Now assume the product converges, with $P_n \to P \neq 0$. We do not claim that the series converges to the principal value of $\log P$, but we will show it converges to some value of $\log P$.

Convergence of the product means that $P_n/P \to 1$, and therefore $\log(P_n/P) \to 0$. Therefore, for each n, there exists and integer h_n such that

$$
\log(P_n/P) = \log(P_n) - \log(P) = [S_n + h_n \cdot 2\pi i] - \log(P). \tag{11.1}
$$

Now looking at differences, we have

$$
\begin{aligned}
(h_{n+1} - h_n)2\pi i &= \log(P_{n+1}/P) - \log(P_n/P) - \log(1 + a_n) \\
(h_{n+1} - h_n)2\pi &= \arg(P_{n+1}/P) - \arg(P_n/P) - \arg(1 + a_n).
\end{aligned}
$$

By assumption $|\arg(1 + a_n)| \leq \pi$. And we know that $\arg(P_{n+1}/P) - \arg(P_n/P) \to 0$. Therefore, for large n and from the above equations, we must have $h_{n+1} = h_n$. Hence, h_n is eventually equal to a fixed integer h. It follows from equation (11.1) that $S_n \to \log P - h \cdot 2\pi i$. So the convergence of the product assures the convergence of the series. $\qquad \square$

11.2 The Jensen Disk

For this section, we define the open disk $D_R = \{z \in \mathbb{C} : |z| < R\}$, with its closure $\overline{D}_R = D_R + \partial D_R$. We also define the open set Ω, with $\overline{D}_R \subset \Omega$. Finally, we assume the complex function $f(z)$ is analytic in Ω. We study here certain properties relating to the zeros of $f(z)$ in D_R.

11.2.1 Jensen's Formula

Lemma 11.8 (Jensen's Formula). *Let $f(z)$ be analytic in Ω with $\overline{D}_R \subset \Omega$, as described above. Suppose $f(0) \neq 0$, $f(z)$ has no zeros on ∂D_R, and $z_1, z_2, ... z_n$ are the zeros of $f(z)$ in D_R (where a zero of order k is included k times in the list). Then*

$$\log \left| f(0) \cdot \frac{R}{z_1} \cdot \frac{R}{z_2} \cdot ... \frac{R}{z_n} \right| = \frac{1}{2\pi} \int_0^{2\pi} \log \left| f(Re^{i\theta}) \right| d\theta.$$

Proof. [32] We first assume that $f(z)$ has no zeros in D_R and apply the Cauchy integral formula:

$$\log(f(z_0)) = \frac{1}{2\pi i} \int_{|z|=R} \frac{\log(f(z))}{z - z_0} dz.$$

Now parameterize the circle with $z = z_0 + Re^{i\theta}$; $0 \le \theta \le 2\pi$:

$$\log(f(z_0)) = \frac{1}{2\pi i} \int_0^{2\pi} \frac{\log\left(f(z_0 + Re^{i\theta})\right)}{(z_0 + Re^{i\theta}) - z_0} \cdot iRe^{i\theta} d\theta$$

$$= \frac{1}{2\pi} \int_0^{2\pi} \log\left(f(z_0 + Re^{i\theta})\right) d\theta.$$

Set $z_0 = 0$ and simplify to:

$$\log(f(0)) = \frac{1}{2\pi} \int_0^{2\pi} \log\left(f(Re^{i\theta})\right) d\theta.$$

But $\log(f(0)) = \log|f(0)| + iArg(f(0))$, so that

$$\log|f(0)| + iArg(f(0)) = \frac{1}{2\pi} \int_0^{2\pi} \left[\log\left| f(Re^{i\theta}) \right| + iArg\left(f(Re^{i\theta}) \right) \right] d\theta$$

$$= \frac{1}{2\pi} \int_0^{2\pi} \log\left| f(Re^{i\theta}) \right| d\theta + i \int_0^{2\pi} \frac{Arg\left(f(Re^{i\theta}) \right)}{2\pi} d\theta.$$

Taking the real parts of both sides, we have (as expected) for the "no zeros" case:

$$\log|f(0)| = \frac{1}{2\pi} \int_0^{2\pi} \log\left| f(Re^{i\theta}) \right| d\theta. \tag{11.2}$$

Now assume $f(z)$ has zeros in D_R and set:

$$F(z) = f(z) \cdot \frac{R^2 - \bar{z}_1 z}{R(z - z_1)} \cdot \frac{R^2 - \bar{z}_2 z}{R(z - z_2)} \cdot ... \frac{R^2 - \bar{z}_n z}{R(z - z_n)}.$$

Of course, each divisor on the RHS removes one of the zeros of $f(z)$. Therefore, $F(z)$ is analytic and has no zeros in D_R, so we can use equation (11.2):

$$\log|F(0)| = \frac{1}{2\pi} \int_0^{2\pi} \log\left| F(Re^{i\theta}) \right| d\theta. \tag{11.3}$$

We note that $\left| \dfrac{R^2 - \bar{z}_j \cdot 0}{R(0 - z_J)} \right| = \left| \dfrac{R}{z_j} \right|$ and thus rewrite the LHS of equation (11.3):

$$\log|F(0)| = \log \left| f(0) \cdot \frac{R^2 - \bar{z}_1 z}{R(z - z_1)} \cdot \frac{R^2 - \bar{z}_2 z}{R(z - z_2)} \cdot ... \frac{R^2 - \bar{z}_n z}{R(z - z_n)} \right|$$

$$= \log \left| f(0) \cdot \frac{R}{z_1} \cdot \frac{R}{z_2} \cdot ... \frac{R}{z_n} \right|.$$

We next note that:

$$\left| \frac{R^2 - \bar{z}_j z}{R(z - z_J)} \right| = 1 \text{ when } |z| = R.$$

To prove this, multiply the numerator by $|\bar{z}|/R$. This does not change the overall modulus because $|z| = R$ and it makes the numerator into the complex conjugate of the denominator.

But that means the RHS of equation (11.3) can be rewritten (replacing $F(z)$ with $f(z)$):

$$\frac{1}{2\pi} \int_0^{2\pi} \log \left| F(Re^{i\theta}) \right| d\theta = \frac{1}{2\pi} \int_0^{2\pi} \left(\log \left| f(Re^{i\theta}) \right| \cdot |1| \right) d\theta.$$

Combining the rewritten LHS and RHS, we have our desired result:

$$\log \left| f(0) \cdot \frac{R}{z_1} \cdot \frac{R}{z_2} \cdot \dots \frac{R}{z_n} \right| = \frac{1}{2\pi} \int_0^{2\pi} \log \left| f(Re^{i\theta}) \right| d\theta. \qquad \square$$

11.2.2 Counting Zeros Inside the Jensen Disk

Define $\mathbf{n}(r)$ as the number of zeros of $f(z)$ (counted with multiplicity) inside the open disk D_r, centered at the origin (unless specifically stated otherwise) and with radius $r > 0$. Clearly, $\mathbf{n}(r)$ is monotone increasing.

Lemma 11.9. *Let $f(z)$ be analytic in Ω with $\overline{D}_R \subset \Omega$, as described above. If $z_1, z_2, \dots z_N$ are the zeros of f inside the disk D_R, then*

$$\int_0^R \mathbf{n}(r) \frac{dr}{r} = \sum_{k=1}^N \log \left| \frac{R}{z_k} \right|.$$

Proof. By the basic rules of logs:

$$\sum_{k=1}^N \log \left| \frac{R}{z_k} \right| = \sum_{k=1}^N \int_{|z_k|}^R \frac{dr}{r}.$$

Now if we define the characteristic function

$$\eta_k(r) = \begin{cases} 1 & \text{if } r > |z_k|, \\ 0 & \text{if } r \le |z_k|, \end{cases}$$

then we have

$$\sum_{k=1}^N \eta_k(r) = \mathbf{n}(r).$$

Therefore

$$\sum_{k=1}^N \log \left| \frac{R}{z_k} \right| = \sum_{k=1}^N \int_{|z_k|}^R \frac{dr}{r} = \sum_{k=1}^N \int_0^R \eta_k(r) \frac{dr}{r} = \int_0^R \left(\sum_{k=1}^N \eta_k(r) \right) \frac{dr}{r} = \int_0^R \mathbf{n}(r) \frac{dr}{r}. \qquad \square$$

Corollary 11.1. *If a function satisfies the requirements of Jensen's Formula, then*

$$\int_0^R \mathbf{n}(r) \frac{dr}{r} = \frac{1}{2\pi} \int_0^{2\pi} \log \left| f(Re^{i\theta}) \right| d\theta - \log |f(0)|.$$

Proof. This follows immediately from lemma 11.9. $\qquad \square$

11.3 The "Order" of Entire Functions

Definition 11.1. *Let $f(z)$ be an entire function. Define*

$$M(r) = max_{|z|=r}|f(z)|.$$

Then $M(r)$ is the maximum value of $f(z)$ on the boundary of a circle of radius r centered at the origin. If the intended function $f(z)$ is not otherwise clear by the context, we will use $M(f, r)$. Note that, by the Maximum Modulus Principle, $M(r)$ is a strictly increasing function of r.

Definition 11.2. *An entire function $f(z)$ is of **finite order** if (and only if) there exists $\lambda, r \in \mathbb{R}_{\geq 0}$ such that*

$$M(r) < e^{|z|^\lambda} \quad \text{for all } |z| > r.$$

*We define ρ as as the **order of growth** of the function f where*

$$\rho = \inf \{\lambda : M(r) < e^{|z|^\lambda} \text{for } |z| \text{ sufficiently large}\},$$

or equally, as the smallest number ρ such that

$$M(r) \leq e^{r^{\rho+\epsilon}} \text{for any given } \epsilon > 0 \text{ as soon as } r \text{ is sufficiently large.}$$

Lemma 11.10. *Let $f(z)$ be an entire function of finite order, and let ρ be the order of growth of $f(z)$. Fix $\epsilon > 0$ and $A \geq 1$. Then*

$$M(r) > Ae^{|z|^{\rho-\epsilon}} \quad \text{for an infinity of values of } r \text{ tending to } +\infty.$$

Proof. We assume there are only a finite number of such r or that such r obtain a maximum, and hope to find a contradiction. In either case, there must be an R greater than the maximum of such r, so that

$$M(r) \leq Ae^{|z|^{\rho-\epsilon}} \quad \text{for all } |z| = r \geq R.$$

We therefore have (for $r \geq R$)

$$M(r) \leq Ae^{|z|^{\rho-\epsilon}} \leq Ae^{r^{\rho-\epsilon}} = Ae^{r^{\rho-\epsilon/2}}e^{r^{-\epsilon/2}} = \left[\frac{A}{e^{r^{\epsilon/2}}}\right]e^{r^{\rho-\epsilon/2}}$$

Now if we set $R' = \max(R, 2/\epsilon \cdot \log A)$, then the bracketed fraction is ≤ 1 and we have

$$M(r) \leq e^{|z|^{\rho-\epsilon/2}} \quad \text{for all } |z| = r \geq R'.$$

We have our contradiction because this would show the order of growth of $f(z) \leq \rho - \epsilon/2$. $\square$

Lemma 11.11. *Let $f(z)$ be an entire function of finite order, and let ρ be the order of growth of $f(z)$. Then*

$$\rho = \limsup_{r \to \infty} \frac{\log \log M(r)}{\log r}.$$

Proof. Fix $\epsilon > 0$. Clearly, there exists an $r_0 > 0$ such that

$$|f(z)| < e^{|z|^{\rho+\epsilon}} \quad \text{for all } |z| > r_0,$$

and therefore (again for some r_0)

$$M(r) < e^{r^{\rho+\epsilon}} \quad \text{for all } r > r_0.$$

From lemma 11.10, setting $A = 1$, we have

$$M(r) > e^{|z|^{\rho-\epsilon}} \quad \text{for an infinity of values of } r \text{ tending to } +\infty.$$

We therefore have

$$\log M(r) < r^{\rho+\epsilon}$$
$$\log \log M(r) < \log r(\rho + \epsilon)$$
$$\frac{\log \log M(r)}{\log r} < \rho + \epsilon,$$

and by the same calculations

$$\frac{\log \log M(r)}{\log r} > \rho - \epsilon \quad \text{for an infinity of values of } r \text{ tending to } +\infty.$$

Thus

$$\rho = \limsup_{r \to \infty} \frac{\log \log M(r)}{\log r}. \qquad \square$$

Lemma 11.12. *Let $f(z)$ be an entire function of finite order ρ. Fix $A, B > 1$. Then ρ is the smallest number such that*
$$M(r) \le Ae^{Br^{\rho+\epsilon}}$$
for any given $\epsilon > 0$ as soon as r is sufficiently large.

Proof. Let λ be the smallest number satisfying our inequality. Because f is of order ρ and $A, B > 1$, we must have that $\lambda \ge \rho$. Our goal, then, is to show that $\lambda \le \rho$. Fix ϵ and set r_0 as the sufficiently large r for the given ϵ. Now set $\delta = 2\epsilon$ and set $r_1 = \max(r_0, 2/\epsilon \cdot \log A, B^{2/\epsilon})$. We have for $r > r_1$:

$$r^{-\epsilon/2} \le \left(B^{2/\epsilon}\right)^{-\epsilon/2} \le 1/B \quad \text{and also} \quad (e^r)^{-\epsilon/2} \le e^{(2/\epsilon \cdot \log A)^{-\epsilon/2}} = e^{-\log A} = 1/A.$$

Thus,

$$M(r) \le Ae^{Br^{\lambda+\epsilon}} = Ae^{B[r^{\lambda+\delta}r^{-\epsilon/2}r^{-\epsilon/2}]} \le Ae^{B[r^{\lambda+\delta}r^{-\epsilon/2}(1/B)]} = Ae^{r^{\lambda+\delta}r^{-\epsilon/2}}$$
$$= Ae^{r^{\lambda+\delta}}e^{r^{-\epsilon/2}} \le Ae^{r^{\lambda+\delta}}e^{(2/\epsilon \cdot \log A)^{-\epsilon/2}} = e^{r^{\lambda+\delta}}.$$

We therefore have that
$$M(r) \le e^{r^{\lambda+\delta}}$$

for any given $\delta > 0$ as soon as r is sufficiently large, and that is only possible if $\lambda \le \rho$, as required. $\square$

Lemma 11.13. *Let $f(z)$ be an entire function of finite order $\leq \rho$, and let $\{z_k\}$ be the zeros of $f(z)$. Then:*

i) $\mathbf{n}(r) \leq Cr^\rho$ *for some $C > 0$ and all sufficiently large r.*

ii) Assume further that all $z_k \neq 0$. Then, for all $s > \rho$ we have

$$\sum_{k=1}^{\infty} \frac{1}{|z_k|^s} < \infty.$$

Proof (i). Assume that $f(0) = 0$ and let $F(z) = f(z)/z^\omega$ where ω is the order of the zero of $f(0)$. Clearly, the order of growth of F is $\leq \rho$. Also, $\mathbf{n}_f(r)$ and $\mathbf{n}_F(r)$ differ only by a constant. That is why, to prove the estimate for $\mathbf{n}(r)$, we may (and will hereafter) assume $f(0) \neq 0$.

For $R > 0$, define the Jensen's Formula integral as

$$JF(R) = \frac{1}{2\pi} \int_0^{2\pi} \log \left| f(Re^{i\theta}) \right| \, d\theta.$$

Our approach will be to: (1) develop an inequality between $\mathbf{n}(r)$ and the Jensen's Formula integral, and (2) develop an inequality between the Jensen's Formula integral and r^ρ.

For our first inequality, set $R = 2r$, recall that $\mathbf{n}(r)$ is non-decreasing and use corollary 11.1:

$$\mathbf{n}(r) \log 2 = \mathbf{n}(r)\left[\log 2r - \log r\right] = \mathbf{n}(r) \int_r^{2r} \frac{dx}{x} \leq \int_r^{2r} \mathbf{n}(x) \frac{dx}{x} \leq \int_0^R \mathbf{n}(x) \frac{dx}{x} = JF(R) - \log|f(0)|.$$

For our second inequality, we use that the order of $f(z)$ is $\leq \rho$ to obtain

$$JF(R) \leq \int_0^{2\pi} \log \left| (e^R)^\rho \right| \, d\theta \leq C' r^\rho$$

for sufficiently large r. Combining our inequalities, we have $\mathbf{n}(r) \leq Cr^\rho$ for an appropriate $C > 0$ and sufficiently large r. $\square$

Proof (i) - Supplemental. If we assume $\rho > \rho_f$, we can eliminate the "sufficiently large r" restriction. Because $f(0) \neq 0$, there is a $\delta_f > 0$ with $f(z) \neq 0$ for $|z| < \delta_f$, and there is an $n_f \in \mathbb{N}$ with $1/n_f < \delta_f$. Fix $r_f = (1/n_f + \delta_f)/2$. Then $\mathbf{n}(r_f) = 0$, so we only need to choose a constant C large enough so that the inequality $\mathbf{n}(r) \leq Cr^\rho$ holds for all $r > r_f$. (Side note: $n_f \cdot r_f > 1$ so that $n_f^\rho \cdot r_f^\rho > 1$). $\square$

Proof (ii). We order our z_k by increasing value of $|z_k|$, and disregard the finite number of $|z_k| < 1$:

$$\sum_{|z_k| \geq 1}^{\infty} |z_k|^{-s} = \sum_{j=0}^{\infty} \left(\sum_{2^j \leq |z_k| < 2^{j+1}} |z_k|^{-s} \right)$$

$$\leq \sum_{j=0}^{\infty} 2^{-js} \mathbf{n}(2^{j+1}) \qquad \text{(because } |z_k| < (2^j)^{-s} = 2^{-js})$$

$$\leq c \sum_{j=0}^{\infty} 2^{-js} 2^{(j+1)\rho} \qquad \text{(because } r < 2^{j+1})$$

$$= c2^{-\rho} \sum_{j=0}^{\infty} (2^{\rho-s})^j < \infty \qquad \text{(which converges because } s > \rho). \qquad \square$$

11.4   Weierstrass Infinite Products

11.4.1   Zero-Free Entire Functions

We know that every non-zero complex number w can be written as $w = e^z$. We would like to extend that idea to non-zero holomorphic functions $f(z)$, with $f(z) = e^{g(z)}$ for some holomorphic function $g(z)$, so that $g(z)$ is one branch of $\log f(z)$.

Lemma 11.14. *If $f(z)$ is a nowhere vanishing holomorphic function in a simply connected region Ω, then there exists a holomorphic function $g(z)$ on Ω such that*

$$f(z) = e^{g(z)}.$$

Proof. [33] Fix a point z_0 in Ω, and define a function

$$g(z) = \int_\gamma \frac{f'(w)}{f(w)}\, dw + w_0,$$

where γ is a path in Ω connecting z_0 to z and w_0 is a complex number so that $e^{w_0} = f(z_0)$. Since f is non-zero and holomorphic, $g(z)$ is also holomorphic with

$$g'(z) = \frac{f'(z)}{f(z)}.$$

We have

$$\frac{d}{dz}\left(f(z)e^{-g(z)}\right) = \frac{d}{dz}\left(\frac{f(z)}{e^{g(z)}}\right) = \frac{f' - g'f}{e^g} = \frac{f' - f'}{e^g} = 0,$$

so that $f(z)e^{-g(z)}$ is constant. With $f(z_0)e^{-w_0} = 1$ and $g(z_0) = w_0$, then $f(z) = e^{g(z)}$ for all $z \in \Omega$. $\square$

11.4.2   Canonical Factors

Weierstrass wanted to construct an entire function that vanishes at all points of a given sequence $\{a_n\}$ and nowhere else. A simple attempt is something of the form

$$\prod_n \left(1 - \frac{z}{a_n}\right).$$

The problem is that convergence of the product depends on the particular sequence $\{a_n\}$. Weierstrass helps to ensure convergence by inserting in the product **canonical factors** (also known as elementary factors), defined for each integer $k \geq 0$ by

$$E_0(z) = (1 - z)$$

$$E_k(z) = (1 - z)e^{S_k(z)} \quad \text{for } k \geq 1 \quad \text{where } S_k(z) = \sum_{n=1}^{k} \frac{z^n}{n}$$

The integer k is called the **degree** of the canonical factor.

We see that $E_k(z/a)$ has a zero at $z = a$ and at no other value of z. What is the idea behind the Weierstrass definition of canonical factors? Note that for $|z| < 1$:

$$-\log(1 - z) = \sum_{n=1}^{\infty} \frac{z^n}{n}$$

so for our $S_k(z)$ defined above

$$\lim_{k \to \infty} S_k(z) = -\log(1 - z)$$

and therefore

$$\lim_{k \to \infty} E_k(z) = (1 - z)e^{-\log(1-z)} = (1 - z)\frac{1}{1 - z} = 1.$$

For increasing k, the value of $E_k(z)$ approaches 1 fairly rapidly, particularly for $|z| \le 1/2$. For example, $E_5(1/2) \approx 0.995$, $E_4(1/4) \approx 0.999$, while $E_{11}(3/4) \approx 0.991$. While not ensuring convergence, use of canonical factors in our infinite products certainly improves chances of convergence. We will discuss the necessary conditions below.

11.4.3 Canonical Factors - Some Inequalities

Lemma 11.15. *Let $z \in \mathbb{C}$. The canonical factors satisfy the following inequalities:*

   i) *If $|z| \le 1/2$, then $|1 - E_k(z)| \le 2e|z|^{k+1}$.*

   ii) *If $|z| \le 1/2$, then $|E_k(z)| \ge e^{-2|z|^{k+1}}$.*

   iii) *If $|z| \ge 1/2$, then $|E_k(z)| \ge |1 - z|e^{-C|z|^k}$.*

Proof. If $k = 0$ then all three results hold, so assume $k \ge 1$. For (i) and (ii) we have $|z| \le 1/2$, so we can use the power series for the logarithm

$$\log(1 - z) = -\sum_{k=1}^{+\infty} \frac{z^k}{k}.$$

When you combine the power series with $(1 - z) = e^{\log(1-z)}$, you have (using S_k is as defined above):

$$E_k(z) = e^{\log(1-z)+S_k} = e^w \quad \text{where} \quad w = -\sum_{n=k+1}^{\infty} \frac{z^n}{n}.$$

For (i), because $|z| < 1/2$, we have

$$|w| \le \sum_{n=k+1}^{\infty} \frac{|z|^n}{n} = \sum_{n=k+1}^{\infty} \frac{|z|^{k+1}|z|^{n-(k+1)}}{n} = |z|^{k+1} \sum_{n=k+1}^{\infty} \frac{|z|^{n-(k+1)}}{n} = |z|^{k+1} \sum_{n=1}^{\infty} \frac{|z|^{n-1}}{n + k}$$

$$\le |z|^{k+1} \left[|z|^0 + \sum_{n=2}^{\infty} \frac{|z|^{n-1}}{n + k} \right] \le |z|^{k+1} \left[1 + \frac{1}{3} \sum_{n=2}^{\infty} \left(\frac{1}{2} \right)^{n-1} \right] < 2|z|^{k+1}.$$

We have shown that $|w| < 1$. Thus:

$$|1 - E_k(z)| = |1 - e^w| = |e^w - 1| = \left| \sum_{n=0}^{\infty} \frac{w^n}{n!} - 1 \right| = \left| \sum_{n=1}^{\infty} \frac{w^n}{n!} \right| \leq \sum_{n=1}^{\infty} \frac{|w|^n}{n!}$$

$$= |w| + \sum_{n=2}^{\infty} \frac{|w|^n}{n!} \leq |w| \left[\sum_{n=0}^{\infty} \frac{1}{n!} \right] = e|w| \leq 2e|z|^{k+1}.$$

This completes (i). Importantly, the desired inequality $|1 - E_k(z)| \leq 2e|z|^{k+1}$ is independent of k.

For (ii), note that $|E_k(z)| = |e^w|$, $|e^w| \geq e^{-|w|}$, and $|w| \leq 2|z|^{k+1}$. The result follows immediately and, again, is independent of k.

For (iii), we start by developing and inequality for $|S_k|$:

$$|S_k| = \left| \sum_{n=1}^{k} \frac{z^n}{n} \right| \leq \sum_{n=1}^{k} \frac{|z|^n}{n} \leq \sum_{n=1}^{k} |z|^n \leq \sum_{n=1}^{k} 2^{k-1}|z|^k \leq k2^{k-1}|z|^k = C|z|^k.$$

We have $|S_k| \leq C|z|^k$ and therefore $e^{-|S_k|} \geq e^{-C|z|^k}$. Because $|e^w| \geq e^{-|w|}$, that means

$$|E_k(z)| = |1 - z||e^{S_k}| \geq |1 - z|e^{-|S_k|} \geq |1 - z|e^{-C|z|^k},$$

as required. $\qquad\square$

We note that, unlike (i) and (ii), in (iii) C is dependent on k. To show that C *must* be dependent on k, set $|z| = 1$ and choose any $C > 0$. We need

$$|S_k| = \left| \sum_{n=1}^{k} \frac{z^n}{n} \right| = \sum_{n=1}^{k} \frac{1}{n} \leq C|z|^k.$$

But the harmonic series diverges, so any fixed C you choose will fail for large enough k. Fortunately, this (iii) is only needed in the *Hadamard Factorization Theorem* where the entire function is of finite order.

11.4.4 Constructing and Entire Functions from Prescribed Zeros

The following theorem from Weierstrass provides a technique for constructing and entire function, based solely on a given sequence of complex numbers. The members of the sequence will be the zeros of the entire function.

Theorem 11.1. *Given any sequence $\{b_n\}$ of complex numbers with $|b_n| \to \infty$ as $n \to \infty$, there exists an entire function f that vanishes at all $z = b_n$ and nowhere else. Any other such entire function is of the form $f(z)e^{g(z)}$, where g is entire.*

Proof. We divide our $\{b_n\}$ into two subsequences, a finite sequence $\{z_n\}$ (described below) and an infinite sequence $\{a_n\}$, with $\{b_n\} = \{z_n\} \cup \{a_n\}$. The finite sequence $\{z_n\}$ consists of those members of $\{b_n\}$ where $b_n = 0$. Let m equal the count (possibly zero) of the members of $\{z_n\}$. Note that $a_n \neq 0$.

As a first attempt, we might define our function $f(z)$ as

$$f(z) = z^m \prod_{n=1}^{\infty} (z - a_n) \quad \text{or} \quad f(z) = z^m \prod_{n=1}^{\infty} \left(1 - \frac{z}{a_n}\right)$$

In both cases, $f(z)$ vanishes at all $z = b_n$ and nowhere else. However, there is no way to be sure that the infinite product converges. We need to solve that problem to create our entire function. The solution is to insert the Weierstrass canonical factors into our product. Specifically, we set

$$f(z) = z^m \prod_{n=1}^{\infty} E_n\left(\frac{z}{a_n}\right) = z^m \prod_{n=1}^{\infty} \left(1 - \frac{z}{a_n}\right) \exp\left[\sum_{k=1}^{n} \frac{1}{n}\left(\frac{z}{a_n}\right)^n\right],$$

where E_n is the canonical factor of degree n.

To see that the product converges for all $z \in \mathbb{C}$, fix R and first verify the result for $|z| < R$. There are only finitely many $|a_n| \le 2R$ (since $|a_n| \to \infty$), so fix N such that $|a_n| > 2R$ for all $n > N$. Our $f(z)$ can then be restated as

$$f(z) = z^m \cdot \prod_{n=1}^{N} E_n\left(\frac{z}{a_n}\right) \cdot \prod_{n=N+1}^{\infty} E_n\left(\frac{z}{a_n}\right)$$

The first product over E_n is finite, vanishes at (and only at) all a_n where $n \le N$, and clearly converges for all z. The second product over E_n is infinite and vanishes at (and only at) all a_n where $n > N$. Regarding convergence of the second product, $|a_n| > 2R$ so that $|z/a_n| \le 1/2$. Applying lemma 11.15 to the second product, we have

$$\left|1 - E_n\left(\frac{z}{a_n}\right)\right| \le 2e \left|\frac{z}{a_n}\right|^{n+1} \le \frac{2e}{2^{n+1}}.$$

We therefore have

$$\sum_{n=N+1}^{\infty} \left|1 - E_n\left(\frac{z}{a_n}\right)\right| = \sum_{n=N+1}^{\infty} \left|E_n\left(\frac{z}{a_n}\right) - 1\right| \le \sum_{n=N+1}^{\infty} \frac{2e}{2^{n+1}} = 2e \sum_{n=1}^{\infty} \frac{1}{2^{n+1}} < 2e.$$

Applying lemma 11.5,

$$\sum_{n=N+1}^{\infty} \left|E_n\left(\frac{z}{a_n}\right) - 1\right| \quad \text{converges} \quad \Rightarrow \quad \prod_{n=N+1}^{\infty} E_n\left(\frac{z}{a_n}\right) \quad \text{converges,}$$

so convergence of $f(z)$ is established, and since R was arbitrarily chosen, our result holds for all R.

Regarding the last statement of the theorem, suppose f_1 and f_2 are both entire functions with zeros given by $\{b_n\}$. Then f_1/f_2 is zero-free and entire. Thus, from lemma 11.14, there exists an entire function g with $f_1/f_2 = e^{g(z)}$. Therefore $f_1(z) = f_2(z)e^{g(z)}$, as required. $\qquad\square$

11.5 Hadamard's Factorization Theorem

11.5.1 Introduction

In Theorem 11.1 (the *Weierstrass Factorization Theorem*), we showed that an entire function that vanishes at $\{b_n\} = \{z_1, z_2, \ldots z_m\} \cup \{a_n\}$, with $z_j = 0$ and $a_j \ne 0$, takes the form

$$e^{g(z)} z^m \prod_{n=1}^{\infty} E_n\left(\frac{z}{a_n}\right).$$

CHAPTER 11. ENTIRE FUNCTIONS

Hadamard's Factorization Theorem is a refinement of the Weierstrass theorem. In the case of functions of finite order, a constant (related to the order of growth) can be used as the degree of the canonical factors, and g is a polynomial.

11.5.2 Preliminary Lemmas

We first establish a lower bound for the product of the canonical factors when z stays away from the zeros $\{a_n\}$, using small disks centered at the zeros.

Lemma 11.16. *Let $f(z)$ be an entire function with order of growth ρ_0, let $k \in \mathbb{N}$ with $k \leq \rho_0 < k+1$, and let $\{a_n\}$ denote the (non-zero) zeros of $f(z)$. For any s with $\rho_0 < s < k+1$, we have*

$$\left| \prod_{n=1}^{\infty} E_k(z/a_n) \right| \geq e^{-c|z|^s},$$

except possibly when z belongs to the union of the disks centered at $\{a_n\}$ of radius $|a_n|^{-k-1}$.

Proof. We split our product into two products, as follows:

$$\prod_{n=1}^{\infty} E_k(z/a_n) = \prod_{|a_n| \leq 2|z|} E_k(z/a_n) \cdot \prod_{|a_n| > 2|z|} E_k(z/a_n).$$

We will call the first product on the RHS *Product A* and the second *Product B*. We will start our proof with *Product B* because it is simpler and still demonstrates our approach to the products.

   Product B. In this case $|z/a_j| < 1/2$ and there are infinitely many such a_j since $a_j \to \infty$. We use lemma 11.15(ii):

$$\left| \prod_{|a_n| > 2|z|} E_k(z/a_n) \right| = \prod_{|a_n| > 2|z|} |E_k(z/a_n)| \geq \prod_{|a_n| > 2|z|} e^{-2|z/a_n|^{k+1}}$$

$$= \prod_{|a_n| > 2|z|} e^{-2|z|^{k+1}|a_n|^{-k-1}} = e^{-2|z|^{k+1} \sum_{|a_n| > 2|z|} \cdot |a_n|^{-k-1}}$$

Now stop to evaluate the sum

$$\sum_{|a_n| > 2|z|} |a_n|^{-k-1} = \sum_{|a_n| > 2|z|} |a_n|^{-s}|a_n|^{s-k-1}.$$

Note that $s < k+1$ and define $\epsilon = k+1-s$ so that $\epsilon > 0$ and $-\epsilon = s-k-1$.

$$= \sum_{|a_n| > 2|z|} |a_n|^{-s}|a_n|^{-\epsilon} = \sum_{|a_n| > 2|z|} |a_n|^{-s} \frac{1}{|a_n|^{\epsilon}}.$$

Next note that $|a_n| > 2|z|$ means $1/(|a_n|)^{\epsilon} < 1/((|2z|)^{\epsilon}$

$$< \sum_{|a_n| > 2|z|} |a_n|^{-s} \frac{1}{|2z|^{\epsilon}} = \frac{1}{|2z|^{\epsilon}} \sum_{|a_n| > 2|z|} |a_n|^{-s} = |2z|^{s-k-1} \sum_{|a_n| > 2|z|} |a_n|^{-s}.$$

Finally note that from lemma 11.13(ii), $\sum |a_n|^{-s}$ converges, so that for some $K > 0$:

$$\sum_{|a_n|>2|z|} |a_n|^{-k-1} < K|2z|^{s-k-1}.$$

Now returning to *Product B*, we have

$$\left| \prod_{|a_n|>2|z|} E_k(z/a_n) \right| \geq e^{-2|z|^{k+1} \sum_{|a_n|>2|z|} |a_n|^{-k-1}} \geq e^{-2|z|^{k+1} K|2z|^{s-k-1}} \geq e^{-c|z|^s}$$

for some $c > 0$, as required.

Product A. In this case $|z/a_j| \geq 1/2$, so there are finitely many such a_j. We use lemma 11.15(iii):

$$\left| \prod_{|a_n|\leq 2|z|} E_k(z/a_n) \right| \geq \prod_{|a_n|\leq 2|z|} \left| 1 - \frac{z}{a_n} \right| \cdot \prod_{|a_n|\leq 2|z|} e^{-C|z/a_n|^k}. \tag{11.4}$$

Looking now at the second product on the RHS, we have

$$\prod_{|a_n|\leq 2|z|} e^{-C|z/a_n|^k} = e^{-C|z|^k \sum_{|a_n|\leq 2|z|} |a_n|^{-k}}.$$

Again, stopping to evaluate the sum

$$\sum_{|a_n|\leq 2|z|} |a_n|^{-k} = \sum_{|a_n|\leq 2|z|} |a_n|^{-s}|a_n|^{s-k}.$$

Note that $0 < (s - k) < 1$ and $|a_n| \leq 2|z|$ means $|a_n|^{s-k} < 2(|z|)^{s-k}$ so that

$$\sum_{|a_n|\leq 2|z|} |a_n|^{-s}|a_n|^{s-k} \leq \sum_{|a_n|\leq 2|z|} |a_n|^{-s}2|z|^{s-k} = 2|z|^{s-k} \sum_{|a_n|\leq 2|z|} |a_n|^{-s}.$$

As above, from lemma 11.13(ii), $\sum |a_n|^{-s}$ converges, so that for some $K' > 0$:

$$\sum_{|a_n|\leq 2|z|} |a_n|^{-k} \leq 2K'|z|^{s-k}.$$

Putting it together, we have

$$\prod_{|a_n|\leq 2|z|} e^{-C|z/a_n|^k} = e^{-C|z|^k \sum_{|a_n|\leq 2|z|} |a_n|^{-k}} \geq e^{-C|z|^k \leq 2K'|z|^{s-k}} = e^{-C'|z|^s},$$

for some $C' > 0$, as required.

Our final task is to estimate the first product on the RHS of equation (11.4). In this case, we will use the assumption that z does not belong to a disk of radius $|a_n|^{-k-1}$ centered at a_n. That is, $|a_n - z| \geq |a_n|^{-k-1}$. Thus

$$\prod_{|a_n|\leq 2|z|} \left| 1 - \frac{z}{a_n} \right| = \prod_{|a_n|\leq 2|z|} \left| \frac{a_n - z}{a_n} \right| \geq \prod_{|a_n|\leq 2|z|} |a_n|^{-k-1}|a_n|^{-1} = \prod_{|a_n|\leq 2|z|} |a_n|^{-k-2}.$$

Using the fact that $z = e^{\log z}$, we can complete the proof if we can show that the logarithm of the last equation above is the required exponential of e. We start with

$$\log\left[\prod_{|a_n|\le 2|z|}|a_n|^{-k-2}\right] = \sum_{|a_n|\le 2|z|}\log\left(\frac{1}{|a_n|^{k+2}}\right) = (k+2)\sum_{|a_n|\le 2|z|}\log\frac{1}{|a_n|} = -(k+2)\sum_{|a_n|\le 2|z|}\log|a_n|.$$

We now stop to evaluate the last sum (without the $-(k+2)$ factor). Because $2|z| \ge |a_j|$, we know the last sum has exactly $\mathbf{n}(2|z|)$ terms. Therefore:

$$\sum_{|a_n|\le 2|z|}\log|a_n| = \sum_{n=1}^{\mathbf{n}(2|z|)}\log|a_n| \le \sum_{n=1}^{\mathbf{n}(2|z|)}\log(2|z|)$$

$$= \mathbf{n}(2|z|)\log(2|z|)$$

$$\le c|z|^s\log(2|z|) \quad \text{From lemma 11.13(i)}$$

$$\le c'|z|^{s'} \quad \text{for any } s' > s.$$

With $s > \rho_0$ and $s' > s$, we can get s' as close to ρ_0 as we like, allowing us to substitute s for s', completing proof. $\qquad\square$

Corollary 11.2. *Continuing with the assumptions of lemma 11.16, we can construct a sequence $\{r_j\}$ of monotone increasing, unbounded and positive real numbers such that*

$$\left|\prod_{n=1}^{\infty}E_k(z/a_n)\right| \ge e^{-c|z|^s} \quad \text{for } |z| = r_j.$$

Proof. We showed in lemma 11.16 that $\sum|a_n|^{-k-1}$ converges. Thus, there is an N such that

$$\sum_{n=N}^{\infty}|a_n|^{-k-1} < \frac{1}{4}.$$

Fix $K > \max(|a_1|, |a_2|, \ldots |a_{N-1}|) + 1$, and define the real interval $K_j = [K + j, K + j + 1]$.

Next, For $n \ge N$, define the real interval

$$I_n = \left[|a_n| - \frac{1}{|a_n|^{k+1}}, |a_n| + \frac{1}{|a_n|^{k+1}}\right],$$

which rotates the lemma 11.16 excluded disk centered at a_n onto the real line.

Now note that, for any given K_j, the union of all I_n cannot fully cover K_j because that would mean $2\sum_{n=N}^{\infty}|a_n|^{-k-1} \ge 1$. Thus, there is an $r_j \in K_j$ such that the circle formed by $|z| = r_j$, does not intersect the excluded disks of lemma 11.16.

We have therefore constructed our required sequence $\{r_j\}$. The corollary follows immediately from lemma 11.16. $\qquad\square$

Lemma 11.17. *Suppose $g(z)$ is an entire function and $u = Re(g(z))$ satisfies*

$$u(z) \le Cr^s \quad \text{whenever } |z| = r$$

for a sequence of positive real numbers r that tends to infinity. Then g is a polynomial of degree $\le s$.

Proof. Because $g(z)$ is an entire function, it can be represented by a power series with an infinite radius of convergence

$$g(z) = \sum_{n=0}^{\infty} a_n (z - z_0)^n \qquad z_0 \in \mathbb{C}.$$

Our proof is complete if we can show that $a_n = 0$ for $n > s$. For convenience, we will take $z_0 = 0$ and center our power series at the origin. Using the Cauchy integral formula, we have

$$a_n = \frac{g^{(n)}(z_0)}{n!} = \frac{1}{2\pi i} \int_{C(r)} \frac{g(\zeta)}{(\zeta - z_0)^{n+1}} d\zeta \qquad n \geq 0, \tag{11.5}$$

where $C(r)$ is a positively oriented circle of radius r centered at z_0. We now parameterize the circle by setting $\zeta = z_0 + re^{i\theta}$, obtaining

$$\begin{aligned} a_n &= \frac{1}{2\pi i} \int_0^{2\pi} \frac{g(z_0 + re^{i\theta})}{(z_0 + re^{i\theta} - z_0)^{n+1}} rie^{i\theta} d\theta \\ &= \frac{1}{2\pi r^n} \int_0^{2\pi} \frac{g(z_0 + re^{i\theta})}{(e^{i\theta})^{n+1}} e^{i\theta} d\theta \\ &= \frac{1}{2\pi r^n} \int_0^{2\pi} g(z_0 + re^{i\theta}) e^{-i(n+1)\theta} e^{i\theta} d\theta \\ &= \frac{1}{2\pi r^n} \int_0^{2\pi} g(z_0 + re^{i\theta}) e^{-in\theta} d\theta. \end{aligned}$$

Although the power series only applies for $n \geq 0$, we can still evaluate the RHS of equation (11.5) for $n < 0$ (i.e., $n \leq -1$ and therefore $-n \geq 1$):

$$\frac{1}{2\pi i} \int_{C(r)} \frac{g(\zeta)}{(\zeta - z_0)^{n+1}} d\zeta = \frac{1}{2\pi r^n} \int_0^{2\pi} g(z_0 + re^{i\theta}) e^{-in\theta} d\theta.$$

We see that the function $g(\zeta)(\zeta - z_0)^{-n-1}$ is entire, so that by Cauchy's theorem the integral on the LHS (and thus the RHS) equals zero. Combining the above results, we have:

$$\frac{1}{2\pi} \int_0^{2\pi} g\left(re^{i\theta}\right) e^{-in\theta} d\theta = \begin{cases} a_n r^n & \text{if } n \geq 0 \\ 0 & \text{if } n < 0. \end{cases} \tag{11.6}$$

We next show that for $n > 0$

$$\frac{1}{2\pi} \int_0^{2\pi} \overline{g\left(re^{i\theta}\right)} e^{-in\theta} d\theta = 0. \tag{11.7}$$

We set $m = -n$, so $m < 0$. We have $\overline{e^{in\theta}} = e^{-in\theta}$, and $\overline{\int g(z) \, dz} = \int \overline{g(z) \, dz}$. Also, note that $\overline{d\theta} = \cos\theta - i\sin\theta$, which simply traverses the circle in the opposite (clockwise) direction:

$$\begin{aligned} \frac{1}{2\pi} \int_0^{2\pi} \overline{g\left(re^{i\theta}\right)} e^{-in\theta} d\theta &= \frac{1}{2\pi} \int_0^{2\pi} \overline{g\left(re^{i\theta}\right) e^{in\theta}} d\theta \\ &= \frac{1}{2\pi} \overline{\int_0^{2\pi} g\left(re^{i\theta}\right) e^{in\theta} \, \overline{d\theta}} \\ &= \frac{1}{2\pi} \overline{\int_0^{2\pi} g\left(re^{i\theta}\right) e^{-im\theta} \, \overline{d\theta}} \\ &= \frac{1}{2\pi} \overline{\int_0^{2\pi} g\left(re^{i\theta}\right) e^{-im\theta} \, d\theta} \qquad \text{(using } m = -n\text{)} \\ &= 0 \qquad \text{(from equation (11.6), since } \overline{0} = 0\text{).} \end{aligned}$$

We can now use the fact that $2u(z) = g(z) + \overline{g(z)}$.

For $n > 0$, we can add equations (11.6) and (11.7) and have

$$a_n r^n = \frac{1}{\pi} \int_0^{2\pi} u\left(re^{i\theta}\right) e^{-in\theta}\, d\theta.$$

For $n = 0$ we can take the real parts of both sides of equation (11.6) and have

$$2Re(a_0) = \frac{1}{\pi} \int_0^{2\pi} u\left(re^{i\theta}\right) d\theta.$$

Using $\oint Cr^s e^{-in\theta}\, d\theta = 0$ for $n \neq 0$, we have

$$a_n = \frac{1}{\pi r^n} \int_0^{2\pi} \left[u\left(re^{i\theta}\right) - Cr^s\right] e^{-in\theta}\, d\theta \quad \text{for } n > 0.$$

We therefore have (separately evaluating the two terms in the integral)

$$|a_n| \leq \frac{1}{\pi r^n} \int_0^{2\pi} \left[Cr^s - u\left(re^{i\theta}\right)\right] d\theta \leq 2Cr^{s-n} - 2Re(a_0)r^{-n}.$$

If we let $r \to \infty$, we see that $a_n = 0$ for $n > s$, as required. $\qquad\square$

11.5.3 Proof of Hadamard's Factorization Theorem

Theorem 11.2 (Hadamard's Factorization Theorem). *Let $f(z)$ be an entire function with order of growth ρ_0, let $k \in \mathbb{N}$ with $k \leq \rho_0 < k + 1$, and let $\{a_n\}$ denote the (non-zero) zeros of $f(z)$. Then*

$$f(z) = e^{P(z)} z^m \prod_{n=1}^{\infty} E_k(z/a_n),$$

where P is a polynomial of degree $\leq k$, and m is the order of the zeros of $f(z)$ at $z = 0$.

Proof. Let

$$h(z) = z^m \prod_{n=1}^{\infty} E_k(z/a_n).$$

where E_k is the canonical factor of degree k.

To prove that $h(z)$ is entire, we must show that the product converges for all $z \in \mathbb{C}$. We use essentially the same argument as in Theorem 11.1.

Fix R and first verify the result for $|z| < R$. There are only finitely many $|a_n| \leq 2R$ (since $|a_n| \to \infty$), so fix N such that $|a_n| > 2R$ for all $n > N$. Our $h(z)$ can then be restated as

$$h(z) = z^m \cdot \prod_{n=1}^{N} E_k\left(\frac{z}{a_n}\right) \cdot \prod_{n=N+1}^{\infty} E_k\left(\frac{z}{a_n}\right)$$

The first product over E_k is finite, vanishes at (and only at) all a_n where $n \leq N$, and clearly converges for all z. The second product over E_k is infinite and vanishes at (and only at) all a_n where $n > N$.

Regarding convergence of the second product, $|a_n| > 2R$ so that $|z/a_n| \leq 1/2$. Applying lemma 11.15 to the second product, we have

$$\left| 1 - E_k\left(\frac{z}{a_n}\right) \right| \leq 2e \left| \frac{z}{a_n} \right|^{k+1} \leq 2eR^{k+1}|a_n|^{-(k+1)}.$$

Let $C = 2eR^{k+1}$. We have

$$\sum_{n=N+1}^{\infty} \left| 1 - E_k\left(\frac{z}{a_n}\right) \right| = \sum_{n=N+1}^{\infty} \left| E_k\left(\frac{z}{a_n}\right) - 1 \right| \leq C \sum_{n=N+1}^{\infty} |a_n|^{-(k+1)}.$$

Because $f(z)$ has order p_0 and $k+1 > p_0$, the last sum above converges by lemma 11.13. Then applying lemma 11.5,

$$\sum_{n=N+1}^{\infty} \left| E_k\left(\frac{z}{a_n}\right) - 1 \right| \quad \text{converges} \quad \Longrightarrow \quad \prod_{n=N+1}^{\infty} E_k\left(\frac{z}{a_n}\right) \quad \text{converges},$$

so convergence of $h(z)$ is established. Since R was arbitrarily chosen, our result holds for all R.

Because $h(z)$ has the zeros of $f(z)$, we know f/h is an entire function with no zeros. Therefore, from lemma 11.14

$$\frac{f(z)}{h(z)} = e^{P(z)} \quad \text{and thus} \quad f(z) = e^{P(z)} h(z)$$

for some entire function P. It only remains to show that $P(z)$ is a polynomial of degree $\leq k$.

Let $\{r_j\}$ represent an ordered sequence of positive real numbers with $r_j \to \infty$ and with $\{r_j\}$ satisfying the criteria of Corollary 11.2. Also, let s be a positive real number such that $k \leq p_0 < s < k + 1$.

By assumption, $f(z)$ has order of growth ρ_0. We use Corollary 11.2 to obtain a minimum value for $h(z)$. Taken together, we have have

$$e^{Re(P(z))} = \left| \frac{f(z)}{h(z)} \right| \leq c' e^{c|z|^s}$$

for $|z| = r_j$. And that means

$$Re(P(z)) \leq C|z|^s \quad \text{for } |z| = r_j.$$

Applying Lemma 11.17, the proof of *Hadamard's Factorization Theorem* is complete. $\qquad \square$

After Riemann's Paper

We have completed our study of Riemann's Paper and have presented our proofs of the (provable) assertions made there. In the remaining chapters in this book, we discuss a selected few of the many developments that followed from Riemann's Paper. With the exception of Chapter 15, we focus on key developments during the period 1895 to 1899.

Chapter 13: Von Mangoldt's Explicit Formula. We provide a proof of von Mangoldt's (improved) version of Riemann's "explicit formula".

Chapter 14: The Prime Number Theorem - A Classic Proof. We provide a proof of the *Prime Number Theorem*. This proof is a slightly simplified version of the original 1896 proofs by Hadamard [18] and de la Vallée Poussin [35].

Chapter 15: Prime Number Theorem - A Modern Proof. We provide a "modern" proof of the *Prime Number Theorem*. This proof follows an approach due to D.J. Newman.

Chapter 16: After the Prime Number Theorem. We discuss developments shortly after proof of the *Prime Number Theorem* relating to the zero-free region for roots of $\zeta(s)$ near $Re(s) = 1$.

In this transition chapter:

Asymptotic Functions. We discuss here asymptotic functions, and attempt to provide clarity on what is meant in the literature by the *error term* of an asymptotic function.

Define Special Functions. We define here several special functions (most importantly the second Chebychev function) that play an important role in the remaining chapters in this book.

Relate $\zeta(s)$ to the Special Functions. The bridge between Riemann's Paper and the special functions defined here is made clear when we provide a proof of the relationship between the logarithmic derivative of $\zeta(s)$ and two of our special functions.

Special Integrals. We provide here proofs of the properties of certain special integrals. Those special integrals are critical to the proofs we develop in the remaining chapters.

12.1   Definitions – Asymptotic Functions

Asymptotic functions are at the center of the *Prime Number Theorem*. In this section we revisit the basics of asymptotic functions, and then discuss the *rate* at which the ratio of two asymptotically equivalent functions converges on 1.

We use $f(x) \sim g(x)$ to indicate that the functions $f(x)$ and $g(x)$ are *asymptotically equivalent*, or equally *asymptotic functions*. They are *asymptotically equivalent* if

$$\lim_{x \to \infty} \frac{f(x)}{g(x)} = 1.$$

That means, for any given $\epsilon > 0$ there is a $K > 0$ such that, for $x > K$, we have

$$(1 - \epsilon) \le \frac{f(x)}{g(x)} \le (1 + \epsilon) \quad \text{or equivalently} \quad \left| \frac{f(x) - g(x)}{g(x)} \right| \le \epsilon.$$

Given two asymptotic functions, you may wish to replace ϵ with $E(x)$, a function of x. Of course, we must have $\lim_{x \to \infty} E(x) = 0$ and we must also have (for sufficiently large x)

$$(1 - E(x)) \le \frac{f(x)}{g(x)} \le (1 + E(x)) \quad \text{and} \quad \left| \frac{f(x) - g(x)}{g(x)} \right| \le E(x).$$

With the above as background, we can now discuss terminology. As $x \to \infty$, the *relative error* between $f(x)$ and $g(x)$ (determined by dividing one by the other) gets increasingly small. So, when $\epsilon = 0.05$, we have $0.95 \le f(x)/g(x) \le 1.05$, so the relative error (for the given x) is $\le 5\%$. And by definition, the relative error for asymptotically equivalent functions must go to zero as $x \to \infty$.

But ϵ tells us nothing about the *rate* at which the relative error approaches zero. For that we need a function. In our example it is $E(x)$. Based on the formula for $E(x)$, we can discover the rate at which $E(x)$ approaches zero, and know that (for large x) the relative error in our asymptotic functions approaches zero *at least as fast* as $E(x)$. We will call $E(x)$ the *error function*.

Our final definition is *error term*. The *error term* tells us the actual (not relative) difference between $f(x)$ and $g(x)$ and is based on the error function. From the equations above we see that the error term (for large x) is

$$|f(x) - g(x)| \le g(x)E(x).$$

It is important to note that our asymptotically equivalent functions do not have just one error term. Any improvement in the function $E(x)$ results in an improved (i.e., reduced) error term. Note also that, unlike our error function, our error term *does not* necessarily (and often does not) go to zero. Consider the asymptotic functions $\pi(x) \sim x/\log x$. In their case, $\lim_{x \to \infty} |\pi(x) - x/\log x| \to \infty$.

We will show in lemma 14.8 that $\psi(x) \sim x$. In our final chapter, we will find an error function $E(x)$ such that

$$\lim_{x \to \infty} E(x) = 0 \quad \text{and (for large enough } x) \quad \left| \frac{\psi(x) - x}{x} \right| \le E(x).$$

Once we find an error function $E(x)$ for our asymptotic functions $\psi(x) \sim x$, we then know the related error term is $xE(x)$.

12.2 Definitions – Special Function

Definition 12.1. $\pi(x)$ *is the* **prime counting function**, *defined for $x > 0$ by*

$$\pi(x) = \sum_{p \le x} 1 \quad \text{for } p \in primes.$$

So, $\pi(x)$ gives a total count of the prime numbers in the interval $[0, x]$.

Definition 12.2. *The **Prime Number Theorem** states that $\pi(x) \sim \dfrac{x}{\log x}$.*

Definition 12.3. *$li(x)$ is the **logarithmic integral**, defined for $\{x \in \mathbb{R}_{\geq 0} : x \neq 1\}$ by*

$$li(x) = \int_0^x \frac{dt}{\log t} \quad or, for\ x > 1 \quad \lim_{\epsilon \to 0+}\left[\int_0^{1-\epsilon} \frac{dt}{\log t} + \int_{1+\epsilon}^x \frac{dt}{\log t} \right].$$

Definition 12.4. *$Li(x)$ is the **offset logarithmic integral**, defined for $x \geq 2$ by*

$$Li(x) = li(x) - li(2) = \int_2^x \frac{dt}{\log t}.$$

Definition 12.5. *$\Lambda(n)$ is the **von Mangoldt function**, defined for $n \in \mathbb{N}$ by*

$$\Lambda(n) = \begin{cases} \log p & \textit{if } n = p^r \textit{ with } p \textit{ a prime number and } r \in \mathbb{N} \\ 0 & \textit{otherwise.} \end{cases}$$

Definition 12.6. *$\theta(x)$ is the **first Chebyshev function** defined for $x > 0$ by*

$$\theta(x) = \sum_{p \leq x} \log p \quad \textit{for all prime numbers } p \leq x.$$

Definition 12.7. *$\psi(x)$ is the **second Chebyshev function**, defined for $x > 0$ by*

$$\psi(x) = \sum_{n \leq x} \Lambda(n) = \sum_{p^n \leq x} \log p = \sum_{p \leq x} \left[\frac{\log x}{\log p} \right] \log p.$$

Here we justify the last equality above for $\psi(x)$. Define a function $r_p(x)$, based on the given prime p and the given x, where $r_p(x)$ is the largest integer such that $p^{r_p(x)} \leq x$. In other words, for a given x, the largest integer r in the definition of $\Lambda(n)$ such that $\Lambda(p^r) = \log p$. For example, for $x = 9$ and $p = 2$, $r_p(x) = 3$ because $2^3 \leq 9$ and $2^4 > 9$. Note that $r_p(x) \cdot \log p$ captures all three $\log p$ terms for the prime number 2 in the formula for $\psi(9)$. For our purposes, we need a different but equivalent expression for $r_p(x)$. We have $r_p(x) = [\log x / \log p]$ where, as usual, $[\ ]$ denotes the greatest integer function.

Definition 12.8. *$\psi_0(x)$ is the **modified second Chebyshev function**, defined for $x > 0$ by*

$$\psi_0(x) = \frac{1}{2}\left(\sum_{n \leq x} \Lambda(n) + \sum_{n < x} \Lambda(n) \right) \begin{cases} \psi(x) - \frac{1}{2}\Lambda(x) & \textit{if } \Lambda(x) \neq 0 \\ \psi(x) & \textit{otherwise.} \end{cases}$$

Throughout the remainder of this book, we will use the above definitions without redefinition.

12.3   Formulae for the Logarithmic Derivative of $\zeta(s)$

We develp here four different formulae for the logarithmic derivative of $\zeta(s)$. These formulae are an important contributor to the advances made after Riemann's Paper.

Lemma 12.1. *For $s \in \mathbb{C}, Re(s) > 1$, we have*

$$(i) \quad -\frac{\zeta'(s)}{\zeta(s)} = \sum_{n=1}^{\infty} \Lambda(n) n^{-s}$$

$$(ii) \quad -\frac{\zeta'(s)}{\zeta(s)} = s \int_{1}^{\infty} \psi(x) x^{-s-1} \, dx$$

$$(iii) \quad -\frac{\zeta'(s)}{\zeta(s)} = \frac{s}{s-1} - \sum_{\rho} \frac{s}{\rho(s-\rho)} + \sum_{n=1}^{\infty} \frac{s}{(2n)(s+2n)} - \frac{\zeta'(0)}{\zeta(0)}$$

$$(iv) \quad -\frac{\zeta'(s)}{\zeta(s)} = \frac{s+1}{2(s-1)} - \sum_{\rho} \frac{s+1}{(\rho+1)(s-\rho)} + \sum_{n=1}^{\infty} \frac{s+1}{(2n-1)(s+2n)} - \frac{\zeta'(-1)}{\zeta(-1)}.$$

Proof (i). We have (where p is an ordered list of primes)

$$\zeta(s) = \prod_{p} \frac{1}{1 - p^{-s}} \quad \text{so that} \quad \log \zeta(s) = -\sum_{p} \log(1 - p^{-s}).$$

Now take the derivative

$$\frac{\zeta'(s)}{\zeta(s)} = -\sum_{p} \frac{p^{-s} \log p}{1 - p^{-s}} = -\sum_{p} \log p \left(\frac{p^{-s}}{1 - p^{-s}} \right)$$

$$-\frac{\zeta'(s)}{\zeta(s)} = \sum_{p} \log p \left(\frac{1}{1 - p^{-s}} - 1 \right) = \sum_{p} \log p \left(\sum_{n=0}^{\infty} p^{-sn} - 1 \right)$$

$$= \sum_{p} \log p \left(\sum_{n=1}^{\infty} p^{-sn} \right) = \sum_{n=1}^{\infty} \Lambda(n) n^{-s}.$$

To justify the last equation, consider the unique factorization of n. If n is 1 or the product of two or more primes (of any multiplicity), then $\Lambda(n) = 0$. Otherwise, $n = p^k$ for some prime p and integer k and $\Lambda(n) = \log p$. $\qquad \square$

Proof (ii). For $n, N \in \mathbb{N}$ we have $\psi(n) - \psi(n-1) = \Lambda(n)$, so that

$$\sum_{n=1}^{N} n^{-s} \Lambda(n) = \sum_{n=1}^{N} n^{-s} [\psi(n) - \psi(n-1)] = \sum_{n=1}^{N} n^{-s} \psi(n) - \sum_{n=1}^{N} n^{-s} \psi(n-1)$$

$$= \sum_{n=1}^{N} n^{-s} \psi(n) - \left[\sum_{n=1}^{1} n^{-s} \psi(n-1) + \sum_{n=2}^{N+1} n^{-s} \psi(n-1) - \sum_{n=N+1}^{N+1} n^{-s} \psi(n-1) \right]$$

$$= \sum_{n=1}^{N} n^{-s} \psi(n) - \left[0 + \sum_{n=1}^{N} (n+1)^{-s} \psi(n) - (N+1)^{-s} \psi(N) \right]$$

$$= \sum_{n=1}^{N} [n^{-s} - (n+1)^{-s}] \psi(n) - (N+1)^{-s} \psi(N).$$

Because $\psi(x) = \mathcal{O}(x)$ and $Re(s) > 1$, the last term is $\mathcal{O}(1/N^{s-1})$ and goes to 0 as $N \to \infty$. We will disregard that term in the intermediate steps below.

$$= \sum_{n=1}^{N} [n^{-s} - (n+1)^{-s}] \psi(n) = \sum_{n=1}^{N} \left[s \int_{n}^{n+1} x^{-s-1} \, dx \right] \psi(n).$$

CHAPTER 12. AFTER RIEMANN'S PAPER

Because $\psi(x)$ is constant in the interval $[n, n+1)$:

$$= \sum_{n=1}^{N} s \int_{n}^{n+1} \psi(x)x^{-s-1}\, dx$$

$$= s \int_{1}^{N} \psi(x)x^{-s-1}\, dx.$$

Taken to the limit, we have

$$s \int_{1}^{\infty} \psi(x)x^{-s-1}\, dx = \lim_{N\to\infty} s \int_{1}^{N} \psi(x)x^{-s-1}\, dx = \lim_{N\to\infty} \sum_{n=1}^{N} n^{-s}\Lambda(n) = \sum_{n=1}^{\infty} \Lambda(n)n^{-s} = -\frac{\zeta'(s)}{\zeta(s)}. \qquad \square$$

Proof (iii). We have from equation (10.7)

$$\log\zeta(s) = \log\xi(0) + \sum_{\rho} \log\left(1 - \frac{s}{\rho}\right) - \log\left(\Gamma\left(\frac{s}{2}+1\right)\right) + \frac{s}{2}\log\pi - \log(s-1),$$

where ρ has its usual meaning (and ordering) as the roots of $\xi(s)$ (see page xi). Now differentiating term-by-term, we have

$$-\frac{\zeta'(s)}{\zeta(s)} = -(\log\zeta(s))' = -\frac{d}{ds}\left[\log\xi(0) + \sum_{\rho}\log\left(1 - \frac{s}{\rho}\right) - \log\left(\Gamma\left(\frac{s}{2}+1\right)\right) + \frac{s}{2}\log\pi - \log(s-1)\right]$$

$$= 0 - \sum_{\rho}\frac{d}{ds}\left(\log\left(1 - \frac{s}{\rho}\right)\right) + \frac{d}{ds}\left(\log\left(\Gamma\left(\frac{s}{2}+1\right)\right)\right) - \frac{1}{2}\log\pi + \frac{d}{ds}\left(\log(s-1)\right).$$

Recall that for $y = \log(f(x))$, we have $dy/dx = f'(x)/f(x)$, so that

$$= 0 - \sum_{\rho}\frac{1}{s-\rho} + \frac{d}{ds}\left(\log\left(\Gamma\left(\frac{s}{2}+1\right)\right)\right) - \frac{1}{2}\log\pi + \frac{1}{s-1}. \qquad (12.1)$$

Rearranging and using the formula for $\log(\Gamma(s))$ from section 10.7

$$= \frac{1}{s-1} - \sum_{\rho}\frac{1}{s-\rho} + \frac{d}{ds}\left(\sum_{n=1}^{\infty}\left[-\log\left(1 + \frac{s}{2n}\right) + \frac{s}{2}\log\left(1 + \frac{1}{n}\right)\right]\right) - \frac{1}{2}\log\pi$$

$$= \frac{1}{s-1} - \sum_{\rho}\frac{1}{s-\rho} + \sum_{n=1}^{\infty}\left[-\frac{1}{s+2n} + \frac{1}{2}\log\left(1 + \frac{1}{n}\right)\right] - \frac{1}{2}\log\pi. \qquad (12.2)$$

Computing for $s = 0$ gives

$$-\frac{\zeta'(0)}{\zeta(0)} = \frac{1}{0-1} - \sum_{\rho}\frac{1}{0-\rho} + \sum_{n=1}^{\infty}\left[-\frac{1}{0+2n} + \frac{1}{2}\log\left(1 + \frac{1}{n}\right)\right] - \frac{1}{2}\log\pi$$

$$= -1 + \sum_{\rho}\frac{1}{\rho} + \sum_{n=1}^{\infty}\left[-\frac{1}{2n} + \frac{1}{2}\log\left(1 + \frac{1}{n}\right)\right] - \frac{1}{2}\log\pi.$$

Subtraction gives

$$-\frac{\zeta'(s)}{\zeta(s)} - \left[-\frac{\zeta'(0)}{\zeta(0)}\right] = \left[\frac{1}{s-1} - \sum_\rho \frac{1}{s-\rho} + \sum_{n=1}^\infty \left[-\frac{1}{s+2n} + \frac{1}{2}\log\left(1+\frac{1}{n}\right)\right] - \frac{1}{2}\log\pi\right]$$

$$-\left[-1 + \sum_\rho \frac{1}{\rho} + \sum_{n=1}^\infty \left[-\frac{1}{2n} + \frac{1}{2}\log\left(1+\frac{1}{n}\right)\right] - \frac{1}{2}\log\pi\right]$$

$$= \left[\frac{1}{s-1} + 1\right] - \sum_\rho \left[\frac{1}{s-\rho} + \frac{1}{\rho}\right] - \sum_{n=1}^\infty \left[\frac{1}{s+2n} - \frac{1}{2n}\right]$$

$$-\frac{\zeta'(s)}{\zeta(s)} + \left[\frac{\zeta'(0)}{\zeta(0)}\right] = \frac{s}{s-1} - \sum_\rho \frac{s}{\rho(s-\rho)} + \sum_{n=1}^\infty \frac{s}{(2n)(s+2n)}$$

$$-\frac{\zeta'(s)}{\zeta(s)} = \frac{s}{s-1} - \sum_\rho \frac{s}{\rho(s-\rho)} + \sum_{n=1}^\infty \frac{s}{(2n)(s+2n)} - \frac{\zeta'(0)}{\zeta(0)}.$$

It remains to justify termwise differentiation of the two infinite series. To do so, we will show that both series in equation (12.2) converge uniformly in any disk $|s| < K$. Beginning with the series that ranges over the roots ρ, we have uniform convergence by lemma 9.8.

Now turning to the series over n, we fix K and have

$$\left|\frac{1}{s+2n} - \frac{1}{2}\log\left(1+\frac{1}{n}\right)\right| = \left|\frac{1}{s+2n} - \frac{1}{2n} + \frac{1}{2n} - \frac{1}{2}\left(\frac{1}{n} - \frac{1}{2n^2} + \frac{1}{3n^3} - \cdots\right)\right|$$

$$\leq \left|\frac{s}{(s+2n)(2n)}\right| + \left|\frac{1}{4n^2} - \frac{1}{6n^3} + \cdots\right| \leq \frac{K}{(2n)^2} + \frac{1}{n^2} = \mathcal{O}\left(\frac{1}{n^2}\right),$$

for sufficiently large n. Since the last converges, the proof is complete. □

Proof (iv). We use (iii) of this lemma and begin with

$$-\frac{\zeta'(s)}{\zeta(s)} = \frac{s}{s-1} - \sum_\rho \frac{s}{\rho(s-\rho)} + \sum_{n=1}^\infty \frac{s}{(2n)(s+2n)} - \frac{\zeta'(0)}{\zeta(0)}.$$

Using this expression, we next add $\zeta'(-1)/\zeta(-1)$ to both sides

$$-\frac{\zeta'(s)}{\zeta(s)} + \frac{\zeta'(-1)}{\zeta(-1)} = \left[\frac{s}{s-1} - \sum_\rho \frac{s}{\rho(s-\rho)} + \sum_{n=1}^\infty \frac{s}{(2n)(s+2n)} - \frac{\zeta'(0)}{\zeta(0)}\right]$$

$$-\left[\frac{-1}{-1-1} - \sum_\rho \frac{-1}{\rho(-1-\rho)} + \sum_{n=1}^\infty \frac{-1}{(2n)(-1+2n)} - \frac{\zeta'(0)}{\zeta(0)}\right]$$

$$= \frac{s+1}{2(s-1)} - \sum_\rho \frac{s+1}{(\rho+1)(s-\rho)} + \sum_{n=1}^\infty \frac{s+1}{(2n-1)(s+2n)}$$

$$-\frac{\zeta'(s)}{\zeta(s)} = \frac{s+1}{2(s-1)} - \sum_\rho \frac{s+1}{(\rho+1)(s-\rho)} + \sum_{n=1}^\infty \frac{s+1}{(2n-1)(s+2n)} - \frac{\zeta'(-1)}{\zeta(-1)}.$$ □

12.4   Some Specialty Integrals

In this section, we introduce three specialty integrals, all well-known in Analytic Number Theory. We will rely on the properties of these integral in several key places in the remaining chapters. For the first two integrals, we follow the proofs in Apostol [2, 243-246].

Lemma 12.2. *Let $z \in \mathbb{C}$, $a > 0$, $c > 0$. Then*

$$\frac{1}{2\pi i} \int_{c-i\infty}^{c+i\infty} a^z \frac{dz}{z} = \begin{cases} 1 & \text{if } a > 1 \\ \frac{1}{2} & \text{if } a = 1 \\ 0 & \text{if } 0 < a < 1, \end{cases}$$

where we define $\int_{c-i\infty}^{c+i\infty}$ to mean $\lim_{T \to \infty} \int_{c-iT}^{c+iT}$ (Cauchy principal value). Also, the following estimates hold

$$\left| \frac{1}{2\pi i} \int_{c-iT}^{c+iT} a^z \frac{dz}{z} - 1 \right| \le \frac{1}{\pi T} \frac{a^c}{\log a} \qquad \text{for } a > 1 \tag{12.3}$$

$$\left| \frac{1}{2\pi i} \int_{c-iT}^{c+iT} a^z \frac{dz}{z} \right| \le \frac{1}{\pi T} \frac{a^c}{\log\left(\frac{1}{a}\right)} \qquad \text{for } a < 1 \tag{12.4}$$

$$\left| \frac{1}{2\pi i} \int_{c-iT}^{c+iT} a^z \frac{dz}{z} - \frac{1}{2} \right| \le \frac{c}{\pi T} \qquad \text{for } a = 1. \tag{12.5}$$

Proof (i). First consider the case where $a > 1$.

Let $b > c > 0$ and $T > c$. Define a contour $C = c + iT \to -b + iT \to -b - iT \to c - iT \to c + iT$, closed and oriented in the positive direction. Define C^- as the same contour, except oriented in the negative direction.

Clearly, $\frac{1}{2\pi i} \int_C a^z \frac{dz}{z} = 1$ because of the simple pole at $z = 0$ with residue 1. So, we have

$$0 = \left[\frac{1}{2\pi i} \int_C a^z \frac{dz}{z} + \frac{1}{2\pi i} \int_{C^-} a^z \frac{dz}{z} \right]$$

$$\frac{1}{2\pi i} \int_{c-iT}^{c+iT} a^z \frac{dz}{z} = \left[\frac{1}{2\pi i} \int_C a^z \frac{dz}{z} + \frac{1}{2\pi i} \int_{C^-} a^z \frac{dz}{z} \right] + \frac{1}{2\pi i} \int_{c-iT}^{c+iT} a^z \frac{dz}{z}$$

$$\frac{1}{2\pi i} \int_{c-iT}^{c+iT} a^z \frac{dz}{z} = 1 + \left(\int_{c-iT}^{-b-iT} + \int_{-b-iT}^{-b+iT} + \int_{-b+iT}^{c+iT} \right) a^z \frac{dz}{z}$$

$$\left| \frac{1}{2\pi i} \int_{c-iT}^{c+iT} a^z \frac{dz}{z} - 1 \right| = \left| \left(\int_{c-iT}^{-b-iT} + \int_{-b-iT}^{-b+iT} + \int_{-b+iT}^{c+iT} \right) a^z \frac{dz}{z} \right|.$$

Our goal is therefore to show that the integral(s) on the RHS go to zero.

For the two horizontal paths, we have

$$\left| \int_{c-iT}^{-b-iT} a^z \frac{dz}{z} \right| \le \int_{-b}^{c} \frac{a^x dx}{\sqrt{T^2}} \le \frac{1}{T} \int_{-\infty}^{c} a^x \, dx = \frac{1}{T} \frac{a^c}{\log a}$$

$$\left| \int_{-b+iT}^{c+iT} a^z \frac{dz}{z} \right| \le \int_{-b}^{c} \frac{a^x dx}{\sqrt{T^2}} \le \frac{1}{T} \int_{-\infty}^{c} a^x \, dx = \frac{1}{T} \frac{a^c}{\log a}.$$

For the vertical path, we see it goes to 0 as $b \to \infty$ (independent of T), so can be disregarded:

$$\left| \int_{-b-iT}^{-b+iT} a^z \frac{dz}{z} \right| \le 2T \frac{a^{-b}}{b}.$$

The first two integrals go to 0 as $T \to \infty$, giving our desired result (including our estimate). $\qquad \square$

Proof (ii). Now consider the case where $0 < a < 1$.

Let $b > c > 0$ and $T > c$. Define a contour $C = b + iT \to c + iT \to c - iT \to b - iT \to b + iT$, closed and oriented in the positive direction. Define C^- as the same contour, except oriented in the negative direction.

Clearly, $\dfrac{1}{2\pi i} \int_C a^z \dfrac{dz}{z} = \dfrac{1}{2\pi i} \int_{C^-} a^z \dfrac{dz}{z} = 0$ because $\dfrac{a^z}{z}$ is analytic in and on C. So we have

$$0 = \left[\frac{1}{2\pi i} \int_{C^-} a^z \frac{dz}{z} \right]$$

$$\frac{1}{2\pi i} \int_{c-iT}^{c+iT} a^z \frac{dz}{z} = \left[\frac{1}{2\pi i} \int_{C^-} a^z \frac{dz}{z} \right] + \frac{1}{2\pi i} \int_{c-iT}^{c+iT} a^z \frac{dz}{z}$$

$$= \left(\int_{c-iT}^{b-iT} + \int_{b-iT}^{b+iT} + \int_{b+iT}^{c+iT} \right) a^z \frac{dz}{z}$$

Our goal is therefore to show that the integral(s) on the RHS go to zero.

For the two horizontal paths, we have (because $0 < a < 1$)

$$\left| \int_{c-iT}^{b-iT} a^z \frac{dz}{z} \right| \le \left| \int_c^b \frac{a^x dx}{\sqrt{T^2}} \right| \le \frac{1}{T} \left| \int_c^\infty a^x \, dx \right| = \frac{1}{T} \left| 0 - \frac{a^c}{\log a} \right| \le \frac{1}{T} \frac{a^c}{\log \left(\frac{1}{a} \right)}$$

$$\left| \int_{b+iT}^{c+iT} a^z \frac{dz}{z} \right| \le \frac{1}{T} \frac{a^c}{\log \left(\frac{1}{a} \right)}.$$

For the vertical path, we see it goes to 0 as $b \to \infty$ (independent of T), so can be disregarded:

$$\left| \int_{b-iT}^{b+iT} a^z \frac{dz}{z} \right| \le 2T \frac{a^b}{b}.$$

The first two integrals go to 0 as $T \to \infty$, giving our desired result (including our estimate). $\quad\square$

Proof (iii). Finally, consider the case where $a = 1$. We have

$$\int_{c-iT}^{c+iT} a^z \frac{dz}{z} = \int_{c-iT}^{c+iT} \frac{dz}{z} = \int_{-T}^T \frac{i}{c + iy} \, dy = \int_{-T}^T \frac{i(c - iy)}{(c - iy)(c + iy)} \, dy = \int_{-T}^T \frac{ic + y}{c^2 + y^2} \, dy$$

$$= \int_{-T}^T \frac{ic}{c^2 + y^2} \, dy + \int_{-T}^T \frac{y}{c^2 + y^2} \, dy$$

$$= \int_{-T}^T \frac{ic}{c^2 + y^2} \, dy + \left[\int_{-T}^0 \frac{y}{c^2 + y^2} \, dy + \int_0^T \frac{y}{c^2 + y^2} \, dy \right].$$

The bracketed integrals cancel, leaving

$$= \int_{-T}^T \frac{ic}{c^2 + y^2} \, dy = 2ic \int_0^T \frac{1}{c^2 + y^2} \, dy.$$

We therefore have

$$\frac{1}{2\pi i} \int_{c-iT}^{c+iT} a^z \frac{dz}{z} = \frac{c}{\pi} \int_0^T \frac{1}{c^2 + y^2} \, dy = \frac{c}{\pi} \left[\frac{1}{c} \arctan \left(\frac{T}{c} \right) \right] = \frac{1}{\pi} \arctan \left(\frac{T}{c} \right)$$

$$= \frac{1}{2} - \frac{1}{\pi} \arctan \left(\frac{c}{T} \right) \le \frac{1}{2} - \frac{c}{\pi T}.$$

The last because $\arctan(c/T) \le c/T$, and we have obtained our desired result (and estimate). $\quad\square$

Theorem 12.1 (Perron's Formula). *Let $s \in \mathbb{C}$, $c > 0$ and $x > 0$. Let $\{f(n)\}$ be an arithmetic function, define*

$$F(s) = \sum_{n=1}^{\infty} \frac{f(n)}{n^s},$$

and assume that $F(s)$ is absolutely convergent for $Re(s) > \sigma$. Then, if $Re(s) > \sigma - c$, we have

$$\frac{1}{2\pi i} \int_{c-i\infty}^{c+i\infty} F(s+z) \frac{x^z}{z}\, dz = \sum_{n<x} \frac{f(n)}{n^s} + \frac{1}{2} \sum_{[[n=x]]} \frac{f(n)}{n^s},$$

where $[[n = x]]$ indicates the term is non-zero only if $n = x$.

Proof. Because $Re(z) = c$ in the integral, and we assume $Re(s) > \sigma - c$, we have $Re(s+z) > \sigma$ and therefore $F(s+z)$ is absolutely convergent. We then have

$$\int_{c-iT}^{c+iT} F(s+z) \frac{x^z}{z}\, dz = \int_{c-iT}^{c+iT} \sum_{n=1}^{\infty} \frac{f(n)}{n^{s+z}} \frac{x^z}{z}\, dz$$

$$= \sum_{n=1}^{\infty} \frac{f(n)}{n^s} \int_{c-iT}^{c+iT} \left(\frac{x}{n}\right)^z \frac{dz}{z}$$

$$= \left(\sum_{n<x} \frac{f(n)}{n^s} + \sum_{[[n=x]]} \frac{f(n)}{n^s} + \sum_{n>x} \frac{f(n)}{n^s} \right) \left(\int_{c-iT}^{c+iT} \left(\frac{x}{n}\right)^z \frac{dz}{z} \right).$$

We now let $T \to \infty$. The first two sums are finite sums, allowing us to apply the limit term by term. For the sum on $n < x$, we have $x/n > 1$, use lemma 12.2 and conclude the integral is $2\pi i$. For the sum on $[[n = x]]$ (assuming x is an integer) we have $x/n = 1$, use lemma 12.2 and conclude the integral is $1/2 \cdot 2\pi i$. We therefore have our result if we can show

$$\lim_{T\to\infty} \sum_{n>x} \frac{f(n)}{n^s} \int_{c-iT}^{c+iT} \left(\frac{x}{n}\right)^z \frac{dz}{z} = 0.$$

We are close to a solution simply by noting $0 < x/n < 1$ and applying lemma 12.2:

$$\sum_{n>x} \frac{f(n)}{n^s} \int_{c-i\infty}^{c+i\infty} \left(\frac{x}{n}\right)^z \frac{dz}{z} = 0.$$

However, because the sum on $n > x$ is an infinite sum, we must also consider the rate that $\lim_{T\to\infty} \to 0$. If we let $a = x/n$ so that $0 < a < 1$, we have from lemma 12.2

$$\left| \int_{c-iT}^{c+iT} a^z \frac{dz}{z} \right| \leq \frac{2}{T} \frac{a^c}{\log\left(\frac{1}{a}\right)}.$$

Now note that $n \geq [x] + 1$, allowing $1/a = n/x \geq ([x]+1)/x$. We therefore have

$$\left| \sum_{n>x} \frac{f(n)}{n^s} \int_{c-iT}^{c+iT} \left(\frac{x}{n}\right)^z \frac{dz}{z} \right| \leq \sum_{n>x} \frac{|f(n)|}{n^\sigma} \left| \frac{2}{T} \left(\frac{x}{n}\right)^c \frac{1}{\log\left(\frac{[x]+1}{x}\right)} \right|$$

$$= \frac{1}{T} \cdot \frac{2x^c}{\log\left(\frac{[x]+1}{x}\right)} \cdot \sum_{n>x} \frac{|f(n)|}{n^{\sigma+c}}.$$

We end up with $1/T$ times a constant (for our fixed x and c) times an absolutely convergent sequence, which therefore must go to 0 as $T \to \infty$. $\qquad\square$

Lemma 12.3. *Let $s, \beta \in \mathbb{C}$, $x > 1$, $c > Re(\beta)$. We have*

$$x^\beta = \frac{1}{2\pi i} \int_{c-i\infty}^{c+i\infty} \frac{x^s}{(s-\beta)} \, ds.$$

Proof. We start with an easily verified formula

$$\int_1^\infty x^{-s} x^{\beta-1} \, dx = \left. \frac{x^{\beta-s}}{\beta-s} \right|_1^\infty = \frac{1}{s-\beta} \qquad \text{for } Re(s-\beta) > 0.$$

With $x > 1$, we can set $x = e^t$ so that $dx = e^t dt$ and our lower bound of integration becomes 0:

$$\frac{1}{s-\beta} = \int_0^\infty e^{-st} e^{t\beta} \, dt.$$

Setting $f(t) = e^{t\beta}$ and $F(s) = 1/(s-\beta)$, we have the form of a Laplace transform

$$\mathscr{L}(f(t)) = F(s) = \frac{1}{s-\beta} = \int_0^\infty e^{-st} e^{t\beta} \, dt = \int_0^\infty e^{-st} f(t) \, dt.$$

That allows the following inverse Laplace transform

$$f(t) = e^{t\beta} = \frac{1}{2\pi i} \int_{c-i\infty}^{c+i\infty} e^{st} \mathscr{L}(f(t)) \, ds = \frac{1}{2\pi i} \int_{c-i\infty}^{c+i\infty} \frac{e^{st}}{s-\beta} \, ds.$$

Now substituting back $e^t = x$, we have our result

$$x^\beta = \frac{1}{2\pi i} \int_{c-i\infty}^{c+i\infty} \frac{x^s}{(s-\beta)} \, ds. \qquad \square$$

Chapter 13

Von Mangoldt's Explicit Formula

13.1 Von Mangoldt's Explicit Formula

Von Mangoldt's approach to developing his explicit formula is similar to Riemann's. However, instead of pairing $\log(\zeta(s))$ with $J(x)$ he pairs $\zeta'(s)/\zeta(s)$ with $\psi(x)$. With the benefit of 37 years hindsight, $\psi(x)$ is a much better choice than Riemann's $J(x)$. For that reason, von Mangoldt's explicit formula has all but replaced Riemann's. We follow Edwards [13, 48-56, 58-61].

As a reminder, the definitions of section 12.2 apply here.

13.1.1 Step 1: Use Perron's Formula

In this step 1, we develop an integral for $\psi_0(x)$ that includes a $\dfrac{\zeta'(s)}{\zeta(s)}$ term. That will be the starting point for step 3, where we use an alternate formula for $\dfrac{\zeta'(s)}{\zeta(s)}$ and obtain our result.

Lemma 13.1. *For $\{s \in \mathbb{C} : Re(s) > 1\}$, $c > 1$, $x > 1$, we have*

$$\psi_0(x) = \frac{1}{2\pi i} \int_{c-i\infty}^{c+i\infty} \left[-\frac{\zeta'(s)}{\zeta(s)} \right] x^s \frac{ds}{s}.$$

Proof. Let $f(n) = \Lambda(n)$ and $F(s) = \sum_{n=1}^{\infty} \Lambda(n)/n^s$. We will first show that $F(s)$ is absolutely convergent for $\sigma = Re(s) > 1$. Let $\epsilon = Re(s) - 1$, so that $\epsilon > 0$. We have

$$|F(s)| = \sum_{n=1}^{\infty} \frac{\Lambda(n)}{|n^s|} = \sum_{n=1}^{\infty} \frac{\Lambda(n)}{n^\sigma} \le \sum_{n=1}^{\infty} \frac{\log n}{n^\sigma} = \sum_{n=1}^{\infty} \left[\frac{1}{n^{1+\epsilon/2}} \cdot \frac{\log n}{n^{\epsilon/2}} \right]$$

Now use lemma 8.1 and choose K such that $\log n < n^{\epsilon/2}$ for $n > K$, giving

$$\le \sum_{n=1}^{K} \left[\frac{1}{n^{1+\epsilon/2}} \cdot \frac{\log n}{n^{\epsilon/2}} \right] + \sum_{n=K+1}^{\infty} \left[\frac{1}{n^{1+\epsilon/2}} \cdot 1 \right] < \infty.$$

Having shown that $F(s)$ is absolutely convergent, we can now use Theorem 12.1 (Perron's Formula):

$$\psi_0(x) = \sum_{n<x} \Lambda(n) + \frac{1}{2} \sum_{[[n=x]]} \Lambda(n) = \frac{1}{2\pi i} \int_{c-i\infty}^{c+i\infty} \left[\sum_{n=1}^{\infty} \frac{\Lambda(n)}{n^s} \right] x^s \frac{ds}{s}.$$

It only remains to substitute for the $F(s)$ term in the integral, using lemma 12.1(i)

$$\psi_0(x) = \frac{1}{2\pi i} \int_{c-i\infty}^{c+i\infty} \left[\frac{\zeta'(s)}{\zeta(s)} \right] x^s \frac{ds}{s}. \qquad \square$$

13.1.2   Step 2: Three Technical Lemmas

The only real difficulty in Step 3 is showing that certain integrals of sums can be evaluated termwise (sums of integrals). We provide the necessary proofs in the three technical lemmas below. Almost all of the difficult work in obtaining von Mangoldt's explicit formula is in lemma 13.4, the last of these three supporting lemmas.

Lemma 13.2. *Let $s \in \mathbb{C}$, $x > 1$, $a > 0$, $d > c \geq 0$. We have*

$$\left| \frac{1}{2\pi i} \int_{a+ic}^{a+id} \frac{x^s}{s} \, ds \right| \leq \frac{x^a}{(a+c)\log x}.$$

Proof. Using integration by parts (with $u(s) = (s \log x)^{-1}$, $v(s) = x^s$)

$$\int \frac{x^s}{s} \, ds = \frac{x^s}{s \log x} + \int \frac{x^s}{s^2 \log x} \, ds,$$

so that

$$\left| \int_{a+ic}^{a+id} \frac{x^s}{s} \, ds \right| \leq \left| \frac{x^{a+id}}{(a+id)\log x} \right| + \left| \frac{x^{a+ic}}{(a+ic)\log x} \right| + \frac{x^a}{\log x} \left| \int_c^d \frac{x^{it}}{(a+it)^2} \, dt \right|.$$

Noting that $|a+id| \geq |a+ic| \geq (a+c)/\sqrt{2}$ (easily seen by letting $a = c + \epsilon$), we have

$$\leq 2 \cdot \frac{\sqrt{2}x^a}{(a+c)\log x} + \frac{x^a}{\log x} \left| \int_c^d \frac{x^{it}}{(a+it)^2} \, dt \right|.$$

We now stop to evaluate the integral immediately above (noting $|x^{it}| = 1$)

$$\left| \int_c^d \frac{x^{it}}{(a+it)^2} \, dt \right| \leq \int_c^\infty \frac{dt}{a^2 + t^2} = \int_0^\infty \frac{dt}{a^2 + (c+t)^2} \leq \int_0^\infty \frac{dt}{a^2 + c^2 + t^2}.$$

Setting $t = u\sqrt{(a^2 + c^2)}$

$$\int_0^\infty \frac{\sqrt{(a^2+c^2)}\, du}{a^2 + c^2 + \left[u\sqrt{(a^2+c^2)} \right]^2} = \frac{1}{\sqrt{(a^2+c^2)}} \int_0^\infty \frac{du}{1+u^2} = \frac{1}{\sqrt{(a^2+c^2)}} \cdot \frac{\pi}{2} \leq \frac{\sqrt{2}}{a+c} \cdot \frac{\pi}{2}.$$

Putting it all together, we have

$$\left| \frac{1}{2\pi i} \int_{a+ic}^{a+id} \frac{x^s}{s} \, ds \right| \leq \frac{1}{2\pi} \left[2 \cdot \frac{\sqrt{2}x^a}{(a+c)\log x} + \frac{\sqrt{2}}{a+c} \cdot \frac{\pi}{2} \right] = \frac{1}{2\pi} \left[\frac{2\sqrt{2}x^a}{(a+c)\log x} + \frac{\sqrt{2}\log x}{(a+c)\log x} \cdot \frac{\pi}{2} \right]$$

$$= \frac{1}{2\pi} \cdot \left(2\sqrt{2} + \frac{\pi}{\sqrt{2}} \right) \cdot \frac{x^a}{(a+c)\log x} \leq \frac{x^a}{(a+c)\log x}. \qquad \square$$

Lemma 13.3. *Fix $c > 1$, fix $x > 1$, let $j \in \{0, 1\}$, and let $k \in \{0, 1\}$. We have*

$$\frac{1}{2\pi i} \int_{c-i\infty}^{c+i\infty} \left[\sum_{n=1}^{\infty} \frac{x^{s+k}}{(2n-j)(s+2n)} \right] ds = \sum_{n=1}^{\infty} \frac{x^k x^{-2n}}{2n-j}. \tag{13.1}$$

Proof. We will temporarily disregard any issues relating to the interchange of sums and integrals and restate the above integral as follows

$$\sum_{n=1}^{\infty} \left[\frac{x^k}{2n-j} \cdot \frac{1}{2\pi i} \int_{c-i\infty}^{c+i\infty} \frac{x^s}{s+2n} \, ds \right]$$

We have $2n > 0$ so that $Re(s - (-2n)) > 0$. Applying lemma 12.3, we have

$$\sum_{n=1}^{\infty} \left[\frac{x^k}{2n-j} \cdot \frac{1}{2\pi i} \int_{c-i\infty}^{c+i\infty} \frac{x^s}{s+2n} \, ds \right] = \sum_{n=1}^{\infty} \frac{x^k x^{-2n}}{2n-j}.$$

It remains to justify the interchange of sum and integral. Clearly, the sum in equation (13.1) converges uniformly in any $|s| < K$. That means the sum/integral interchange is permitted over any *finite* interval. Our goal, then, is to take the finite intervals to the limit and show that the limit of the sum is equal to the sum of the limit.

For the sum of the limit, we again apply lemma 12.3 and have

$$\sum_{n=1}^{\infty} \left[\frac{x^k}{2n-j} \left(\lim_{T\to\infty} \frac{1}{2\pi i} \int_{c-iT}^{c+iT} \frac{x^s}{s+2n} \, ds \right) \right] = \sum_{n=1}^{\infty} \left[\frac{x^k}{2n-j} \left(x^{-2n} \right) \right] = \sum_{n=1}^{\infty} \frac{x^k x^{-2n}}{2n-j},$$

and we have our convergence. For the limit of the sum, we start with

$$\lim_{T\to\infty} \sum_{n=1}^{\infty} \left[\frac{x^k}{2n-j} \cdot \frac{1}{2\pi i} \int_{c-iT}^{c+iT} \frac{x^s}{s+2n} \, ds \right] = \lim_{T\to\infty} \sum_{n=1}^{\infty} \left[\frac{x^k x^{-2n}}{2n-j} \cdot \frac{1}{2\pi i} \int_{c-iT}^{c+iT} \frac{x^{s+2n}}{s+2n} \, ds \right].$$

Disregarding the finite number of terms where $n^2 < c$, we evaluate the nth term in the series. Set $t = s + 2n$ so that (using lemma 13.2)

$$= \frac{x^k x^{-2n}}{2n-j} \cdot \left| \frac{1}{2\pi i} \int_{c+2n-iT}^{c+2n+iT} \frac{x^t}{t} \, dt \right| = \frac{x^k x^{-2n}}{2n-j} \cdot 2 \left| \frac{1}{2\pi i} \int_{c+2n-i0}^{c+2n+iT} \frac{x^t}{t} \, dt \right|$$

$$\leq \frac{x^k x^{-2n}}{2n-j} \cdot 2 \cdot \frac{x^{c+2n}}{(c+2n)\log x} = \frac{2}{2n-j} \cdot \frac{x^{c+k}}{(c+2n)\log x} \leq \frac{2x^{c+k}}{\log x} \cdot \frac{1}{4n^2 - jc} \leq \frac{2x^{c+k}}{\log x} \cdot \frac{1}{n^2},$$

valid for all T. We have uniform convergence of the series, justifying the interchange of sum and integral. $\square$

Lemma 13.4. *Fix $c > 1$, fix $x > 1$, let $j \in \{0, 1\}$, and let $k \in \{0, 1\}$. We have*

$$\frac{1}{2\pi i} \int_{c-i\infty}^{c+i\infty} \left[\sum_{\rho} \frac{x^k x^s}{(\rho+j)(s-\rho)} \right] ds = \sum_{\rho} \frac{x^k x^{\rho}}{(\rho+j)}.$$

Proof. For our fixed x, the x^k term is like a constant and can be moved outside the integral, giving

$$\frac{1}{2\pi i} \int_{c-i\infty}^{c+i\infty} \left[\sum_{\rho} \frac{x^k x^s}{(\rho+j)(s-\rho)} \right] ds = x^k \cdot \frac{1}{2\pi i} \int_{c-i\infty}^{c+i\infty} \left[\sum_{\rho} \frac{x^s}{(\rho+j)(s-\rho)} \right] ds.$$

We will not consider the x^k term in the remainder of this proof. Thus, we only need to prove

$$\frac{1}{2\pi i} \int_{c-i\infty}^{c+i\infty} \left[\sum_{\rho} \frac{x^s}{(\rho+j)(s-\rho)} \right] ds = \sum_{\rho} \frac{x^{\rho}}{(\rho+j)}. \tag{13.2}$$

We will temporarily disregard any issues relating to the interchange of sums and integrals and restate equation (13.2) as follows

$$= \sum_\rho \left[\frac{1}{(\rho + j)} \cdot \frac{1}{2\pi i} \int_{c-i\infty}^{c+i\infty} \frac{x^s}{s - \rho} \, ds \right]. \tag{13.3}$$

We have $0 \le Re(\rho) \le 1$ so that $Re(s - \rho) > 0$. Applying lemma 12.3, we have

$$= \sum_\rho \frac{x^\rho}{(\rho + j)}.$$

It remains to justify the interchange of sum and integral.

We first show that the infinite series inside the brackets of equation (13.2) converges uniformly in $|s| < K$ for any fixed $K > 0$. We have

$$\left| \sum_\rho \frac{x^s}{(\rho + j)(s - \rho)} \right| \le x^c \left| \sum_\rho \frac{1}{(\rho + j)(s - \rho)} \right| \le x^c \sum_\rho \left| \frac{1}{(\rho + j)} \right| \left| \frac{1}{(s - \rho)} \right|.$$

Disregarding the finite number of terms where $|(\rho + j)| < 1$, we have

$$\le x^c \sum_\rho \left| \frac{1}{(s - \rho)} \right|.$$

We apply lemma 9.8 and have uniform convergence in $|s| < K$.

That means the sum/integral interchange in equation (13.3) is permitted over any *finite* interval.

Our goal, then, is to take the finite intervals to the limit and show that the limit of the sum is equal to the sum of the limit.

For the sum of the limit we have by lemma 12.3

$$\sum_\rho \left[\frac{1}{(\rho + j)} \cdot \frac{1}{2\pi i} \lim_{T \to \infty} \int_{c-iT}^{c+iT} \frac{x^s}{s - \rho} \, ds \right] = \sum_\rho \frac{x^\rho}{(\rho + j)}. \tag{13.4}$$

For the limit of the sum we have

$$\lim_{T \to \infty} \cdot \frac{1}{2\pi i} \int_{c-iT}^{c+iT} \left[\sum_\rho \frac{x^s}{(\rho + j)(s - \rho)} \right] ds,$$

where we have shown above that the sum over ρ converges over any finite interval. We obtain our result, therefore, if we can show the RHS of equation (13.4) converges. This is the most difficult step in von Mangoldt's proof. We start with a slightly modified form of the LHS of equation (13.4)

$$\lim_{T \to \infty} \sum_{|Im(\rho)| \le T} \left[\frac{x^\rho}{(\rho + j)} \cdot \frac{1}{2\pi i} \int_{c-iT}^{c+iT} \frac{x^{s-\rho}}{s - \rho} \, ds \right].$$

Using that form of equation, we wish to show that $A(T)$ and $B(T)$ go to 0 as $T \to \infty$:

$$A(T) = \sum_\rho \left[\frac{x^\rho}{(\rho + j)} \cdot \frac{1}{2\pi i} \int_{c-iT}^{c+iT} \frac{x^{s-\rho}}{s - \rho} \, ds \right] - \sum_{|Im(\rho)| \le T} \left[\frac{x^\rho}{(\rho + j)} \cdot \frac{1}{2\pi i} \int_{c-iT}^{c+iT} \frac{x^{s-\rho}}{s - \rho} \, ds \right]$$

$$B(T) = \sum_{|Im(\rho)| \le T} \left[\frac{x^\rho}{(\rho + j)} \cdot \frac{1}{2\pi i} \int_{c-iT}^{c+iT} \frac{x^{s-\rho}}{s - \rho} \, ds \right] - \sum_{|Im(\rho)| \le T} \frac{x^\rho}{(\rho + j)}.$$

Beginning with $A(T)$, let $\rho = \beta + i\gamma$ represent the form of a given root ρ. We have

$$|A(T)| = \sum_{|\gamma|>T} \left| \frac{x^\rho}{(\rho+j)} \right| \cdot \left| \frac{1}{2\pi i} \int_{c-iT}^{c+iT} \frac{x^{s-\rho}}{s-\rho} \, ds \right|$$

$$\leq 2 \sum_{\gamma>T} \frac{x^\beta}{\gamma} \cdot \left| \frac{1}{2\pi i} \int_{c-\beta+i(\gamma-T)}^{c-\beta+i(\gamma+T)} \frac{x^t}{t} \, dt \right|$$

$$\leq 2 \sum_{\gamma>T} \frac{x^\beta}{\gamma} \cdot \frac{x^{c-\beta}}{(c-\beta+\gamma-T)\log x} \qquad \text{(using lemma 13.2)}$$

$$\leq 2 \frac{x^c}{\log x} \cdot \sum_{\gamma>T} \frac{1}{\gamma(c-\beta+\gamma-T)}.$$

We have reduced $A(T)$ to a constant times a sum, so our goal is to show that the sum $\to 0$ as $T \to \infty$. Let $n \in \mathbb{N}_{\geq 0}$ and let I_n be the interval $(T+n, T+n+1]$. Then we have

$$\sum_{\gamma>T} \frac{1}{\gamma(c-\beta+\gamma-T)} = \sum_{n=0}^{\infty} \sum_{\gamma\in I_n} \frac{1}{\gamma(c-\beta+\gamma-T)}.$$

We would like to continue our inequality by expressing the sum over I_n as a single value (based on n). In what follows, we assume that T is large enough so that the estimate of lemma 9.6 applies. Therefore the nth such interval contains no more than $2\log(T+n)$ of the γ. That allows the numerator of our single value in I_n to equal $2\log(T+n)$. We need a denominator for I_n that: (1) does not include γ, and (2) for each γ in the interval, is less than $\gamma(c-\beta+\gamma-T)$.

Recall that $c > 1$ and $0 < \beta < 1$. Let $a = c-1$ so that $a > 0$ and $a \leq c-\beta$ for all ρ. We then have in any interval I_n

$$\gamma(c-\beta+\gamma-T) \geq \gamma(a+\gamma-T) \geq \gamma(a+n) \geq (T+n)(a+n). \qquad (13.5)$$

Now returning to our full sum, and using the inequality in equation (13.5) above, we have

$$\sum_{\gamma>T} \frac{1}{\gamma(c-\beta+\gamma-T)} = \sum_{n=0}^{\infty} \sum_{\gamma\in I_n} \frac{1}{\gamma(c-\beta+\gamma-T)}$$

$$\leq \sum_{n=0}^{\infty} \frac{2\log(T+n)}{(T+n)(a+n)} = 2 \sum_{n=0}^{\infty} \frac{\log(T+n)}{(T+n)(a+n)}$$

Next, choose T large enough such that, for all $n \geq 0$, $\log(T+n) < \sqrt{T+n}$.

$$\leq 2 \sum_{n=0}^{\infty} \left(\frac{\log(T+n)}{(T+n)(a+n)} \cdot \frac{\sqrt{T+n}}{\log(T+n)} \right) = 2 \sum_{n=0}^{\infty} \frac{1}{\sqrt{T+n} \cdot (a+n)}.$$

Now observe that $\sqrt{T+n} \cdot (a+n) \geq (T+n)^{1/4}(a+n)^{1/4}(a+n) \geq T^{1/4}n^{5/4}$

$$\leq 2 \sum_{n=0}^{\infty} \frac{1}{T^{1/4}n^{5/4}} = \frac{2}{T^{1/4}} \sum_{n=0}^{\infty} \frac{1}{n^{5/4}},$$

and we have reduced the sum to a constant times $1/T^{1/4}$ which $\to 0$ as $T \to \infty$, as required.

Finally, we turn to $B(T)$ and wish to show $B(T) \to 0$ as $T \to \infty$. We start with

$$|B(T)| = \sum_{|Im(\rho)|\leq T} \left| \frac{x^\rho}{(\rho + k)} \right| \cdot \left| \frac{1}{2\pi i} \int_{c-iT}^{c+iT} \frac{x^{s-\rho}}{s - \rho} ds - 1 \right|$$

$$= 2 \sum_{0<\gamma\leq T} \left| \frac{x^\rho}{(\rho + k)} \right| \cdot \left| \frac{1}{2\pi i} \int_{c-iT}^{c+iT} \frac{x^{s-\rho}}{s - \rho} ds - 1 \right|.$$

Symmetry allows use of either $x^{s-\rho}$ or $x^{s-\overline{\rho}}$. Since we assume $\gamma > 0$, then $\overline{\rho}$ has the form $\beta - i\gamma$.

$$= 2 \sum_{0<\gamma\leq T} \left| \frac{x^{\overline{\rho}}}{(\overline{\rho} + k)} \right| \cdot \left| \frac{1}{2\pi i} \int_{c-iT}^{c+iT} \frac{x^{s-\overline{\rho}}}{s - \overline{\rho}} ds - 1 \right|$$

$$\leq 2 \sum_{0<\gamma\leq T} \frac{x^\beta}{\gamma} \cdot \left| \frac{1}{2\pi i} \int_{(c-\beta)-i(T+\gamma)}^{(c-\beta)+i(T-\gamma)} \frac{x^t}{t} dt - 1 \right|$$

$$= 2 \sum_{0<\gamma\leq T} \frac{x^\beta}{\gamma} \cdot \left| \frac{1}{2\pi i} \left(\int_{(c-\beta)-i(T+\gamma)}^{(c-\beta)+i(T+\gamma)} - \int_{(c-\beta)+i(T-\gamma)}^{(c-\beta)+i(T+\gamma)} \right) \frac{x^t}{t} dt - 1 \right|$$

$$\leq 2 \sum_{0<\gamma\leq T} \frac{x^\beta}{\gamma} \cdot \left| \frac{1}{2\pi i} \int_{(c-\beta)-i(T+\gamma)}^{(c-\beta)+i(T+\gamma)} \frac{x^t}{t} dt - 1 \right| + 2 \sum_{0<\gamma\leq T} \frac{x^\beta}{\gamma} \cdot \left| \frac{1}{2\pi i} \int_{(c-\beta)+i(T-\gamma)}^{(c-\beta)+i(T+\gamma)} \frac{x^t}{t} dt \right|.$$

We can now use our estimates from lemma 12.2, equation (12.3) for the first integral and lemma 13.2 for the second integral:

$$\leq 2 \sum_{0<\gamma\leq T} \frac{x^\beta}{\gamma} \cdot \frac{x^{c-\beta}}{\pi (T + \gamma) \log x} + 2 \sum_{0<\gamma\leq T} \frac{x^\beta}{\gamma} \cdot \frac{x^{c-\beta}}{(c - \beta + T - \gamma) \log x}.$$

With the above, if we let $a = c - 1$ (as was used in the $A(T)$ case), we can conclude that

$$|B(T)| \leq \frac{2x^c}{\pi \log x} \sum_{0<\gamma\leq T} \frac{1}{\gamma (T + \gamma)} + \frac{2x^c}{\log x} \sum_{0<\gamma\leq T} \frac{1}{\gamma (a + T - \gamma)}. \tag{13.6}$$

To show that $|B(T)| \to 0$, we must show that the two sums in equation (13.6) go to 0 as $T \to \infty$.

Let $T, T', k \in \mathbb{N}$. Assume $T > T'$ and assume T' is large enough so that the estimate of lemma 9.6 applies. Let I_k' be the interval $(T' + k, T' + k + 1]$.

We consider first the left sum in equation (13.6). We have

$$\sum_{0<\gamma\leq T} \frac{1}{\gamma (T + \gamma)} \leq \sum_{\gamma\leq T'} \frac{1}{\gamma (T + \gamma)} + \sum_{k=0}^{T-T'-1} \sum_{\gamma\in I_k'} \frac{1}{\gamma (T + \gamma)}$$

$$\leq \sum_{\gamma\leq T'} \frac{1}{\gamma (T + \gamma)} + \sum_{k=0}^{T-T'-1} \frac{2 \log(T' + k)}{(T' + k)(T + T' + k)}.$$

By Theorem 9.1, the first sum has no more than $A \log(T')$ terms for some constant A and thus goes to 0 as $T \to \infty$. For the second sum

$$\sum_{k=0}^{T-T'-1} \frac{2 \log(T' + k)}{(T' + k)(T + T' + k)} \leq \sum_{k=0}^{T-T'-1} \frac{2 \log T}{T} \cdot \frac{1}{T' + k} = \frac{2 \log T}{T} \sum_{k=0}^{T-T'-1} \frac{1}{T' + k}$$

$$\leq \frac{2 \log T}{T} \int_{T'-1}^{T} \frac{dt}{t} \leq \frac{2(\log T)^2}{T},$$

which also goes to 0 as $T \to \infty$.

Finally, consider the right sum in equation (13.6). For the reasons discussed above, the finite sum over the terms with $\gamma \leq T'$ goes to 0 as $T \to \infty$. Thus, only the following sum remains

$$= \sum_{k=0}^{T-T'-1} \sum_{\gamma \in I_k'} \frac{1}{\gamma(a+T-\gamma)} \leq \sum_{k=0}^{T-T'-1} \frac{2\log(T'+k)}{(T'+k)(a+T-T'-k-1)}$$

$$\leq 2\log T \sum_{k=0}^{T-T'-1} \frac{1}{(a+T-1)} \left(\frac{1}{T'+k} + \frac{1}{a+T-T'-k-1} \right)$$

$$\leq \frac{2\log T}{(T-1)} \sum_{k=0}^{T-T'-1} \frac{1}{T'+k} + \frac{2\log T}{(T-1)} \sum_{k=0}^{T-T'-1} \frac{1}{a+T-T'-k-1}.$$

The left sum was previously shown to go to 0 (we can disregard the -1 term in the denominator). For the right sum, we have

$$= \frac{2\log T}{(T-1)} \left[\sum_{k=0}^{T-T'-2} \frac{1}{a+T-T'-k-1} + \sum_{k=T-T'-1}^{T-T'-1} \frac{1}{a+T-T'-k-1} \right]$$

$$= \frac{2\log T}{(T-1)} \sum_{k=0}^{T-T'-2} \frac{1}{a+T-T'-k-1} + \frac{2\log T}{(T-1)} \cdot \frac{1}{a}$$

$$\leq \frac{2\log T}{(T-1)} \cdot \sum_{k=1}^{T} \frac{1}{k} + \frac{2\log T}{(T-1)} \cdot \frac{1}{a} \leq \frac{2\log T}{(T-1)} \cdot \left[(1+\log T) + \frac{1}{a} \right],$$

and the last goes to 0 as $T \to \infty$, as required. $\qquad \square$

13.1.3 Step 3: Evaluate Integral Term-by-Term

With the work of Steps 1 and 2 complete, we can almost coast to the finish line.

Theorem 13.1 (von Mangoldt's Explicit Formula). *For $x > 1$, we have*

$$\boxed{\psi_0(x) = x - \sum_{\rho} \frac{x^\rho}{\rho} + \sum_{n=1}^{\infty} \frac{x^{-2n}}{2n} - \frac{\zeta'(0)}{\zeta(0)} \qquad (x > 1).}$$

Proof. We start with the integral developed in lemma 13.1.

$$\psi_0(x) = \frac{1}{2\pi i} \int_{c-i\infty}^{c+i\infty} \left[\frac{\zeta'(s)}{\zeta(s)} \right] x^s \frac{ds}{s}.$$

Next, we use lemma 12.1(iii) and substitute for the bracketed term

$$\psi_0(x) = \frac{1}{2\pi i} \int_{c-i\infty}^{c+i\infty} \left[\frac{s}{s-1} - \sum_{\rho} \frac{s}{\rho(s-\rho)} + \sum_{n=1}^{\infty} \frac{s}{(2n)(s+2n)} - \frac{\zeta'(0)}{\zeta(0)} \right] x^s \frac{ds}{s}. \qquad (13.7)$$

Expressed as separate integrals (term by term) we have

$$\psi_0(x) = \frac{1}{2\pi i} \int_{c-i\infty}^{c+i\infty} \frac{x^s}{s-1} \, ds - \frac{1}{2\pi i} \int_{c-i\infty}^{c+i\infty} \left[\sum_{\rho} \frac{x^s}{\rho(s-\rho)} \right] ds$$

$$+ \frac{1}{2\pi i} \int_{c-i\infty}^{c+i\infty} \left[\sum_{n=1}^{\infty} \frac{x^s}{(2n)(s+2n)} \right] ds - \frac{\zeta'(0)}{\zeta(0)} \left[\frac{1}{2\pi i} \int_{c-i\infty}^{c+i\infty} \frac{x^s}{s} \, ds \right]. \qquad (13.8)$$

Now consider the four terms on the RHS individually. For the first term, we have $Re(s-1) > 0$ and can therefore apply lemma 12.3 to have

$$\frac{1}{2\pi i}\int_{c-i\infty}^{c+i\infty}\frac{x^s}{s-1}\,ds = x^1 = x. \tag{13.9}$$

For the second term, we apply lemma 13.4 (setting $j = k = 0$) and have

$$-\frac{1}{2\pi i}\int_{c-i\infty}^{c+i\infty}\left[\sum_\rho\frac{x^s}{\rho(s-\rho)}\right]ds = -\sum_\rho\frac{x^\rho}{\rho}. \tag{13.10}$$

For the third term, we apply lemma 13.3 (setting $j = k = 0$) and have

$$\sum_{n=1}^{\infty}\left[\frac{1}{2n}\cdot\frac{1}{2\pi i}\int_{c-i\infty}^{c+i\infty}\frac{x^s}{s+2n}\,ds\right] = \sum_{n=1}^{\infty}\frac{x^{-2n}}{2n}. \tag{13.11}$$

For the forth (and final) term, we have $x > 1$ and can apply lemma 12.2 to have

$$-\frac{\zeta'(0)}{\zeta(0)}\left[\frac{1}{2\pi i}\int_{c-i\infty}^{c+i\infty}\frac{x^s}{s}\,ds\right] = -\frac{\zeta'(0)}{\zeta(0)}\cdot 1 = -\frac{\zeta'(0)}{\zeta(0)}. \tag{13.12}$$

Combining equations (13.9), (13.10), (13.11) and (13.12), we have von Mangoldt's final result:

$$\psi_0(x) = x - \sum_\rho\frac{x^\rho}{\rho} + \sum_{n=1}^{\infty}\frac{x^{-2n}}{2n} - \frac{\zeta'(0)}{\zeta(0)} \qquad (x > 1). \qquad\qquad \square$$

13.2 A Final Comment

As a point of interest, von Mangoldt was able to evaluate the constant $\zeta'(0)/\zeta(0) = \log 2\pi$. Also, by using the series expansion

$$\log\left[\frac{1}{(1-x)}\right] = x + \frac{x^2}{2} + \frac{x^3}{3} + \cdots \qquad (x > 1),$$

we can evaluate the next to last term

$$\sum_{n=1}^{\infty}\frac{x^{-2n}}{2n} = \frac{1}{2}\sum_{n=1}^{\infty}\frac{x^{-2^n}}{n} = \frac{1}{2}\log\left[\frac{1}{(1-x^{-2})}\right] = \frac{1}{2}\log\left[\frac{x^2}{(x^2-1)}\right] \qquad (x > 1).$$

This allows another useful form of Von Mangoldt's final formula

$$\psi_0(x) = x - \sum_\rho\frac{x^\rho}{\rho} + \frac{1}{2}\log\left[\frac{x^2}{(x^2-1)}\right] - \log 2\pi \qquad (x > 1).$$

Chapter 14

The Prime Number Theorem - A Classic Proof

14.1 Our Approach

The *Prime Number Theorem* was (simultaneously) proved 37 years after Riemann's Paper by Hadamard [18] and de la Vallée Poussin [35]. This proof is a slightly simplified version of their approach. We generally follow Edwards [13, 68-69, 72-77], and prove the *Prime Number Theorem* in seven steps, as outlined below.

Step 1 – Definitions We define certain functions that may be used in the proof.
Step 2 – No Zeros We prove that $\zeta(s) \neq 0$ for $Re(s) = 1$.
Step 3 – Develop an Integral We develop an alternate integral formula for $\int_1^x \psi(t)\,dt$.
Step 4 – Integral to Formula Using our integral formula we develop another formula for $\int_1^x \psi(t)\,dt$.
Step 5 – Asymptotic Equivalence Using our last formula, we show that $\int_1^x \psi(t)\,dt \sim x^2/x$.
Step 6 – Equivalent Statement We show $\psi(x) \sim x$ if and only if $\pi(x) \sim x/\log x$.
Step 7 – The Finish We show $\int_1^x \psi(t)\,dt \sim x^2/x \Longrightarrow \psi(x) \sim x$, proving the *Prime Number Theorem*.

14.2 Step 1 – Definitions

We incorporate here all of the definitions included in section 12.2. This includes the definition of the *Prime Number Theorem*: $\pi(x) \sim x/\log x$.

14.3 Step 2 – No Zeros

We begin this section by explaining[34] the importance of proving that $\zeta(s) \neq 0$ for $Re(s) = 1$.

We will show in lemma 14.6 that the *Prime Number Theorem* is proved if and only if $\psi(x) \sim x$, or equally, $\lim_{x\to\infty} \psi(x)/x = 1$. Thus

$$\lim_{x\to\infty} \frac{\psi(x)}{x} = 1 \iff \lim_{x\to\infty} \frac{\psi(x)}{x} - 1 = 0 \iff \lim_{x\to\infty} \frac{\psi(x)}{x} - \frac{x}{x} = 0 \iff \lim_{x\to\infty} \frac{\psi(x) - x}{x} = 0.$$

Combining the last equation with equation (13.10), we have that $\psi(x) \sim x$ if and only if

$$\lim_{x\to\infty} \frac{-\sum_\rho \frac{x^\rho}{\rho} + \frac{1}{2} \log\left[\frac{x^2}{(x^2-1)}\right] - \log 2\pi}{x} = 0.$$

The last term in the numerator is a constant. And it is easily seen that the next to last term in the numerator is monotone decreasing and rapidly approaches zero for increasing x. Thus both terms may be disregarded as $x \to \infty$. That means $\psi(x) \sim x$ if and only if

$$\lim_{x \to \infty} \frac{\sum_\rho \frac{x^\rho}{\rho}}{x} = \lim_{x \to \infty} \sum_\rho \frac{x^{\rho-1}}{\rho} = 0.$$

Looking at any one term of the last sum, we need to show that $\lim_{x \to \infty} x^{\rho-1} = 0$. Clearly, that requires $Re(\rho) < 1$. We already know that $\zeta(s) \neq 0$ for $Re(s) > 1$. Therefore, if we show that the last sum can be evaluated termwise, proving the *Prime Number Theorem* boils down to proving that $\zeta(s) \neq 0$ for $Re(s) = 1$. The following very clever proof is due to Mertens [28].

Lemma 14.1. *For all $\theta \in \mathbb{R}$, we have $3 + 4\cos\theta + \cos 2\theta \geq 0$.*

Proof. Using the identity $\cos^2\theta = [1 + \cos 2\theta]/2$, we have

$$3 + 4\cos\theta + \cos 2\theta = 3 + 4\cos\theta + 2\cos^2\theta - 1 = 2(1 + \cos\theta)^2 \geq 0. \qquad \square$$

Lemma 14.2. *Let $s = x + iy \in \mathbb{C}$. Then, $\zeta(s) \neq 0$ on the line $Re(s) = 1$.*

Proof.[35] We begin by evaluating $\log|\zeta(s)|$ for $Re(s) > 1$:

$$\log|\zeta(s)| = \log\left|\prod_p (1 - p^{-s})^{-1}\right| = -\sum_p \log|(1 - p^{-s})| = -Re \sum_p \log(1 - p^{-s})$$

The last because $Re(\log(z)) = Re\left[\log|z| + iArgz\right] = \log|z|$.

$$= Re \sum_p -\log(1 - p^{-s}) = Re\left[\sum_p \sum_{n=1}^\infty \frac{1}{n}p^{-ns}\right].$$

The last using the Taylor Series: $-\log(1 - z) = \sum_{n=1}^\infty \frac{x^n}{n}$ for $|z| < 1$.

Now fix $y \neq 0$ and define a function:

$$h(x) = \zeta^3(x)\zeta^4(x + iy)\zeta(x + i2y).$$

Therefore

$$\log|h(x)| = 3\log|\zeta(x)| + 4\log|\zeta(x + iy)| + \log|\zeta(x + i2y)|$$

$$= 3Re\left[\sum_p \sum_{n=1}^\infty \frac{1}{n}p^{-nx}\right] + 4Re\left[\sum_p \sum_{n=1}^\infty \frac{1}{n}p^{-nx-iny}\right] + Re\left[\sum_p \sum_{n=1}^\infty \frac{1}{n}p^{-nx-i2nt}\right]$$

$$= Re\left[\sum_p \sum_{n=1}^\infty \frac{1}{n}p^{-nx}\left(3 + 4p^{-iny} + p^{-i2ny}\right)\right]$$

$$= \sum_p \sum_{n=1}^\infty \frac{1}{n}p^{-nx}Re\left[3 + 4p^{-iny} + p^{-i2ny}\right]$$

$$= \sum_p \sum_{n=1}^\infty \frac{1}{n}p^{-nx}Re\left[3 + 4e^{-iny\log p} + e^{-i2ny\log p}\right].$$

For our fixed y and any given p and n, define $\theta_{p,n} = -ny \log p$, so that

$$= \sum_{p} \sum_{n=1}^{\infty} \frac{1}{n} p^{-nx} Re \left[3 + 4 \left(\cos \theta_{p,n} + i \sin \theta_{p,n} \right) + \left(\cos 2\theta_{p,n} + i \sin 2\theta_{p,n} \right) \right]$$

$$= \sum_{p} \sum_{n=1}^{\infty} \frac{1}{n} p^{-nx} \left[3 + 4 \cos \theta_{p,n} + \cos 2\theta_{p,n} \right].$$

Applying lemma 14.1 to the last equation above, we see that $\log |h(x)| \geq 0$ and therefore

$$|h(x)| = |\zeta^3(x)||\zeta^4(x + iy)||\zeta(x + i2y)| \geq 1.$$

We can rewrite the last inequality and have

$$|(x - 1)\zeta(x)|^3 \cdot \left| \frac{\zeta(x + iy)}{(x - 1)} \right|^4 \cdot |\zeta(x + i2y)| \geq \frac{1}{(x - 1)}. \tag{14.1}$$

We now assume $\zeta(1 + iy) = 0$ and hope to obtain a contradiction. For each of the three terms on the LHS of equation (14.1), consider the limit as $x \to 1+$.

For the first term, $\zeta(s)$ is holomorphic on $\mathbb{C} \setminus [1]$, with a simple pole at $s = 1$ with residue 1. Thus, $\lim_{x \to 1} (x - 1)\zeta(x) = 1$. More generally, for $s \in \mathbb{C}$ we have that $(s - 1)\zeta(s)$ is holomorphic on $\mathbb{C}$ and therefore bounded in any neighborhood of $s = 1$.

For the second term, we know $y \neq 0$ because of the simple pole of $\zeta(s)$ at 1. Thus, $\zeta(1 + iy)$ is analytic in a neighborhood of $1 + iy$. We can use L'Hôpital's rule and have

$$\lim_{x \to 1+} \frac{\zeta(x + iy)}{(x - 1)} = \lim_{x \to 1+} \frac{d/dx \ \zeta(x + iy)}{d/dx \ (x - 1)} = \lim_{x \to 1+} \zeta'(x + iy),$$

which is likewise analytic (and therefore bounded) in a neighborhood of $1 + iy$.

For the third term, we know $2y \neq 0$. Therefore, this term is analytic (and therefore bounded) in a neighborhood of $1 + i2y$.

Putting it together, we have

$$\lim_{x \to 1+} |(x - 1)\zeta(x)|^3 \cdot \left| \frac{\zeta(x + iy)}{(x - 1)} \right|^4 \cdot |\zeta(x + i2y)| \geq \lim_{x \to 1+} \frac{1}{(x - 1)}.$$

We have our contradiction because the LHS of the inequality is bounded and the RHS is unbounded. We conclude that $\zeta(1 + iy) \neq 0$ for $y \neq 0$, and therefore ζ has no zeros on $Re(s) = 1$. $\qquad \square$

14.4 Step 3 – Develop an Integral

Lemma 14.3. *For $s \in \mathbb{C}$, $c > 1$, $x > 1$,*

$$\int_1^x \psi(t) \, dt = \frac{1}{2\pi i} \int_{c - i\infty}^{c + i\infty} \left[-\frac{\zeta'(s)}{\zeta(s)} \right] \frac{x^{s+1}}{s(s + 1)} \, ds.$$

Proof. We start by defining $F(x)$

$$F(x) = \frac{1}{2\pi i} \int_{c - i\infty}^{c + i\infty} \left[-\frac{\zeta'(s)}{\zeta(s)} \right] \frac{x^{s+1}}{s(s + 1)} \, ds.$$

Apply lemma 12.1(i), substituting for the bracketed term. Subject to justifying termwise integration, we have

$$F(x) = \frac{1}{2\pi i} \int_{c-i\infty}^{c+i\infty} \left[\sum_{n=1}^{\infty} \frac{\Lambda(n)}{n^s} \right] \frac{x^{s+1}}{s(s+1)} \, ds = \sum_{n=1}^{\infty} \Lambda(n) \frac{1}{2\pi i} \int_{c-i\infty}^{c+i\infty} \frac{x^{s+1}}{n^s s(s+1)} \, ds.$$

Now using $\frac{1}{s(s+1)} = \frac{1}{s} - \frac{1}{(s+1)}$, we have

$$= \sum_{n=1}^{\infty} \Lambda(n) \frac{1}{2\pi i} \int_{c-i\infty}^{c+i\infty} \frac{x^{s+1}}{n^s s} \, ds - \sum_{n=1}^{\infty} \Lambda(n) \frac{1}{2\pi i} \int_{c-i\infty}^{c+i\infty} \frac{x^{s+1}}{n^s (s+1)} \, ds$$

$$= \sum_{n=1}^{\infty} \Lambda(n) \frac{x}{2\pi i} \int_{c-i\infty}^{c+i\infty} \left(\frac{x}{n} \right)^s \frac{ds}{s} - \sum_{n=1}^{\infty} \Lambda(n) \frac{1}{2\pi i} \int_{c-i\infty}^{c+i\infty} \frac{n}{s+1} \left(\frac{x}{n} \right)^{s+1} \, ds.$$

Letting $u = s + 1$ in the second integral gives

$$= \sum_{n=1}^{\infty} \Lambda(n) \frac{x}{2\pi i} \int_{c-i\infty}^{c+i\infty} \left(\frac{x}{n} \right)^s \frac{ds}{s} - \sum_{n=1}^{\infty} \Lambda(n) \frac{n}{2\pi i} \int_{c+1-i\infty}^{c+1+i\infty} \left(\frac{x}{n} \right)^u \frac{du}{u}.$$

Now use lemma 12.2 and the fact that $x - n = 0$ when $x = n$:

$$= \sum_{n \le x} \Lambda(n)(x - n) = \sum_{n \le x} [\psi(n) - \psi(n-1)] (x - n).$$

With telescoping we have

$$= \sum_{n=1}^{[x]-1} \psi(n) + \psi([x])(x - [x]).$$

Because $\psi(x)$ is constant in the interval $[n, n+1)$:

$$= \sum_{n=1}^{[x]-1} \int_n^{n+1} \psi(t) \, dt + \int_{[x]}^x \psi(t) \, dt$$

$$= \int_1^x \psi(t) \, dt.$$

We therefore have

$$F(x) = \int_1^x \psi(t) \, dt = \frac{1}{2\pi i} \int_{c-i\infty}^{c+i\infty} \left[-\frac{\zeta'(s)}{\zeta(s)} \right] \frac{x^{s+1}}{s(s+1)} \, ds,$$

so our proof is complete if we can justify termwise integration of the following two integrals

$$\frac{1}{2\pi i} \int_{c-i\infty}^{c+i\infty} x \sum_{n=1}^{\infty} \Lambda(n) \left(\frac{x}{n} \right)^s \frac{ds}{s} \quad \text{and} \quad \frac{1}{2\pi i} \int_{c+1-i\infty}^{c+1+i\infty} \sum_{n=1}^{\infty} n\Lambda(n) \left(\frac{x}{n} \right)^u \frac{du}{u}.$$

For the sum of the limit, we again use lemma 12.2 (see page xi for the $[[n = x]]$ definition)

$$\sum_{n=1}^{\infty} x\Lambda(n) \lim_{T \to \infty} \frac{1}{2\pi i} \int_{c-iT}^{c+iT} \left(\frac{x}{n} \right)^s \frac{ds}{s} = \sum_{n<x} x\Lambda(n) + \frac{1}{2} \sum_{[[n=x]]} x\Lambda(n)$$

$$\sum_{n=1}^{\infty} n\Lambda(n) \lim_{T \to \infty} \frac{1}{2\pi i} \int_{c+1-iT}^{c+1+iT} \left(\frac{x}{n} \right)^u \frac{du}{u} = \sum_{n<x} n\Lambda(n) + \frac{1}{2} \sum_{[[n=x]]} n\Lambda(n),$$

with convergence in both cases.

For the limit of the sum, we start with the first integral

$$\lim_{T \to \infty} \sum_{n=1}^{\infty} \left[x \Lambda(n) \frac{1}{2\pi i} \int_{c-iT}^{c+iT} \left(\frac{x}{n} \right)^s \frac{ds}{s} \right]$$

and evaluate the nth term in the series. We exclude the first finite number of terms and begin with the first n such that $\log n > 2 \log x$ (and thus $n > x$), using the estimate of lemma 12.2

$$\leq x \Lambda(n) \cdot \frac{1}{\pi T} \cdot \frac{(\frac{x}{n})^c}{\log(n/x)} = \frac{x^{c+1}}{\pi T} \cdot \frac{\Lambda(n)}{n^c \log(n/x)} \leq \frac{x^{c+1}}{\pi T} \cdot \frac{\log n}{n^c \log(n/x)} \leq \frac{x^{c+1}}{\pi T} \cdot \frac{2}{n^c}.$$

Because $c > 1$, we have uniform convergence of the series, justifying the exchange of sum and integral.

Now consider the limit of the sum for the second integral

$$\lim_{T \to \infty} \sum_{n=1}^{\infty} \left[n \Lambda(n) \frac{1}{2\pi i} \int_{c+1-iT}^{c+1+iT} \left(\frac{x}{n} \right)^u \frac{du}{u} \right]$$

and evaluate the nth term in the series. We exclude the first finite number of terms and begin with the first n such that $\log n > 2 \log x$ (and thus $n > x$)

$$\leq n \Lambda(n) \cdot \frac{1}{\pi T} \cdot \frac{(\frac{x}{n})^{c+1}}{\log(n/x)} = \frac{x^{c+1}}{\pi T} \cdot \frac{n \Lambda(n)}{n^{c+1} \log(n/x)} \leq \frac{x^{c+1}}{\pi T} \cdot \frac{n \log n}{n^{c+1} \log(n/x)} \leq \frac{x^{c+1}}{\pi T} \cdot \frac{2}{n^c}.$$

Because $c > 1$, we have uniform convergence of the series, justifying the exchange of sum and integral. This completes the proof. $\qquad \square$

14.5 Step 4 – Integral to Formula

Lemma 14.4. *Let ρ have its usual meaning (and ordering) as the roots of $\xi(s)$ (see page xi). and let $x > 1$. We have*

$$\int_1^x \psi(t) \, dt = \frac{x^2}{2} - \sum_{\rho} \frac{x^{\rho+1}}{\rho(\rho+1)} - \sum_{n=1}^{\infty} \frac{x^{1-2n}}{2n(2n-1)} - x \frac{\zeta'(0)}{\zeta(0)} + \frac{\zeta'(-1)}{\zeta(-1)}.$$

Proof. We begin with the integral from lemma 14.3 and again use $\dfrac{1}{s(s+1)} = \dfrac{1}{s} - \dfrac{1}{(s+1)}$

$$\int_1^x \psi(t) \, dt = \frac{1}{2\pi i} \int_{c-i\infty}^{c+i\infty} \left[-\frac{\zeta'(s)}{\zeta(s)} \right] \frac{x^{s+1}}{s(s+1)} \, ds \quad (x > 1, \ c > 1)$$

$$= \frac{1}{2\pi i} \int_{c-i\infty}^{c+i\infty} \left[-\frac{\zeta'(s)}{\zeta(s)} \right] x^{s+1} \left(\frac{1}{s} - \frac{1}{s+1} \right) \, ds.$$

Our integrand is therefore equal to

$$\left[-\frac{\zeta'(s)}{\zeta(s)} \right] x^{s+1} \left(\frac{1}{s} - \frac{1}{s+1} \right) = \left[-\frac{\zeta'(s)}{\zeta(s)} \right] \frac{x^{s+1}}{s} - \left[-\frac{\zeta'(s)}{\zeta(s)} \right] \frac{x^{s+1}}{s+1}.$$

We use the two expressions for $-\zeta'(s)/\zeta(s)$ in lemma 12.1(iii) & (iv) to obtain as our integrand

$$= \left[\frac{s}{s-1} - \sum_\rho \frac{s}{\rho(s-\rho)} + \sum_{n=1}^\infty \frac{s}{(2n)(s+2n)} - \frac{\zeta'(0)}{\zeta(0)} \right] \frac{x^{s+1}}{s}$$

$$- \left[\frac{s+1}{2(s-1)} - \sum_\rho \frac{s+1}{(\rho+1)(s-\rho)} + \sum_{n=1}^\infty \frac{s+1}{(2n-1)(s+2n)} - \frac{\zeta'(-1)}{\zeta(-1)} \right] \frac{x^{s+1}}{s+1}$$

$$= \left[\frac{x^{s+1}}{s-1} - \sum_\rho \frac{x^{s+1}}{\rho(s-\rho)} + \sum_{n=1}^\infty \frac{x^{s+1}}{(2n)(s+2n)} - \frac{\zeta'(0)}{\zeta(0)} \frac{x^{s+1}}{s} \right]$$

$$- \left[\frac{x^{s+1}}{2(s-1)} - \sum_\rho \frac{x^{s+1}}{(\rho+1)(s-\rho)} + \sum_{n=1}^\infty \frac{x^{s+1}}{(2n-1)(s+2n)} - \frac{\zeta'(-1)}{\zeta(-1)} \frac{x^{s+1}}{s+1} \right].$$

For the four terms that are not sums over ρ or n, we slightly rewrite them

$$= \frac{x^{s+1}}{s-1} - \frac{x^{s+1}}{2(s-1)} - \frac{\zeta'(0)}{\zeta(0)} \frac{x^{s+1}}{s} + \frac{\zeta'(-1)}{\zeta(-1)} \frac{x^{s+1}}{s+1}$$

$$= \frac{x^2}{2} \frac{x^{s-1}}{(s-1)} - x \frac{\zeta'(0)}{\zeta(0)} \frac{x^s}{s} + \frac{\zeta'(-1)}{\zeta(-1)} \frac{x^{s+1}}{s+1},$$

and see that, applying lemma 12.2, the integral for those four terms resolves to

$$\boxed{\frac{x^2}{2} - x\frac{\zeta'(0)}{\zeta(0)} + \frac{\zeta'(-1)}{\zeta(-1)}}.$$

We now consider the two sums over ρ. We apply lemma 13.4 with $j = k = 1$ in the first integral and $j = 0$, $k = 1$ in the second integral

$$= \frac{1}{2\pi i} \int_{c-i\infty}^{c+i\infty} \left[\sum_\rho \frac{x^{s+1}}{(\rho+1)(s-\rho)} \right] ds - \frac{1}{2\pi i} \int_{c-i\infty}^{c+i\infty} \left[\sum_\rho \frac{x^{s+1}}{\rho(s-\rho)} \right] ds$$

$$= \boxed{\sum_\rho \frac{x^{\rho+1}}{(\rho+1)} - \sum_\rho \frac{x^{\rho+1}}{\rho}}.$$

Finally, we consider the two sums over n.

$$= \frac{1}{2\pi i} \int_{c-i\infty}^{c+i\infty} \left[\sum_{n=1}^\infty \frac{x^{s+1}}{(2n)(s+2n)} \right] ds - \frac{1}{2\pi i} \int_{c-i\infty}^{c+i\infty} \left[\sum_{n=1}^\infty \frac{x^{s+1}}{(2n-1)(s+2n)} \right] ds.$$

Now apply lemma 13.3, with $j = 0$, $k = 1$ in the first integral and $j = k = 1$ in the second:

$$= \boxed{\sum_{n=1}^\infty \frac{x^{1-2n}}{2n} - \sum_{n=1}^\infty \frac{x^{1-2n}}{(2n-1)}}.$$

Putting it all together, we have

$$\int_1^x \psi(t)\, dt = \frac{x^2}{2} - \sum_\rho \frac{x^{\rho+1}}{\rho} + \sum_\rho \frac{x^{\rho+1}}{(\rho+1)} + \sum_{n=1}^\infty \frac{x^{1-2n}}{2n} - \sum_{n=1}^\infty \frac{x^{1-2n}}{(2n-1)} - x\frac{\zeta'(0)}{\zeta(0)} + \frac{\zeta'(-1)}{\zeta(-1)}$$

$$= \frac{x^2}{2} - \sum_\rho \frac{x^{\rho+1}}{\rho(\rho+1)} - \sum_{n=1}^\infty \frac{x^{1-2n}}{2n(2n-1)} - x\frac{\zeta'(0)}{\zeta(0)} + \frac{\zeta'(-1)}{\zeta(-1)}. \qquad \square$$

14.6   Step 5 – Asymptotic Equivalence

Lemma 14.5. *The following asymptotically equivalent relationship holds:* $\int_0^\infty \psi(t)\, dt \sim \dfrac{x^2}{2}$.

Proof. To prove $\int_0^\infty \psi(t)\, dt \sim \dfrac{x^2}{2}$, we must show: $\lim\limits_{x \to \infty} \left| \dfrac{\int_0^x \psi(t)\, dt - \frac{x^2}{2}}{\frac{x^2}{2}} \right| = 0$.

We start with the results of lemma 14.4 and have

$$\int_0^x \psi(t)\, dt - \frac{x^2}{2} = -\sum_\rho \frac{x^{\rho+1}}{\rho(\rho+1)} - \sum_{n=1}^\infty \frac{x^{-2n+1}}{2n(2n-1)} - x\frac{\zeta'(0)}{\zeta(0)} + \frac{\zeta'(-1)}{\zeta(-1)}. \qquad (14.2)$$

Dividing both sides of equation (14.2) by $x^2/2$ and evaluating the limit as $x \to \infty$, it is immediately clear by inspection that all terms on the RHS can be disregarded except for the sum over ρ. That is

$$\lim_{x \to \infty} \left| \frac{\int_0^x \psi(t)\, dt - \frac{x^2}{2}}{\frac{x^2}{2}} \right| = \lim_{x \to \infty} \left| \sum_\rho \frac{x^{\rho-1}}{\rho(\rho+1)} \right|. \qquad (14.3)$$

We have from lemma 9.7 that $\sum_\rho [\rho(\rho+1)]^{-1}$ converges absolutely. By lemma 14.2, $|x^{\rho-1}| < 1$. Thus, the full series in equation (14.3) converges uniformly. That allows us to evaluate the limit termwise. Because $\lim_{x \to \infty}$ of each term goes to 0, the limit of the full series goes to 0, as required. $\square$

14.7   Step 6 – Equivalent Statement

Lemma 14.6. *We have* $\psi(x) \sim x$ *if and only if* $\pi(x) \sim x/\log x$.

Proof.[36] We start with

$$\psi(x) = \sum_{p \le x} \left[\frac{\log x}{\log p} \right] \log p \le \sum_{p \le x} \frac{\log x}{\log p} \log p = \log x \sum_{p \le x} 1 = \log x\, \pi(x). \qquad (14.4)$$

Now let $1 < y < x$, so that

$$\pi(x) = \pi(y) + \sum_{y < p \le x} 1$$

$$\le \pi(y) + \sum_{y < p \le x} \frac{\log p}{\log(y)} \quad \text{(since } p > y\text{)}$$

$$= \pi(y) + \frac{1}{\log(y)} \sum_{y < p \le x} \log p \le y + \frac{1}{\log(y)} \sum_{y < p \le x} \log p \le y + \frac{1}{\log(y)} \psi(x).$$

Using the above result, set $y = x/(\log x)^2$ to obtain

$$\pi(x) \le \frac{x}{(\log x)^2} + \frac{1}{\log x - 2\log(\log x)} \psi(x),$$

so that

$$\pi(x)\frac{\log x}{x} \le \frac{1}{\log x} + \frac{\log x}{\log x - 2\log(\log x)}\frac{\psi(x)}{x}. \tag{14.5}$$

Therefore, combining equation (14.4) with equation (14.5)

$$\frac{\psi(x)}{x} \le \pi(x)\frac{\log x}{x} \le \frac{1}{\log x} + \frac{\log x}{\log x - 2\log(\log x)}\frac{\psi(x)}{x}.$$

Now assume that $\psi(x) \sim x$

$$\lim_{x\to\infty}\frac{\psi(x)}{x} = 1 \le \lim_{x\to\infty}\left[\pi(x)\frac{\log x}{x}\right] \le \lim_{x\to\infty}\left[\frac{1}{\log x} + \frac{\log x}{\log x - 2\log(\log x)}\right] = 1,$$

giving $\pi(x) \sim x/\log x$, as required. Finally assume $\pi(x) \sim x/\log x$

$$\lim_{x\to\infty}\frac{\psi(x)}{x} \le \lim_{x\to\infty}\left[\pi(x)\frac{\log x}{x}\right] = 1 \le \lim_{x\to\infty}\left[\frac{1}{\log x} + \frac{\log x}{\log x - 2\log(\log x)}\frac{\psi(x)}{x}\right] = \lim_{x\to\infty}\frac{\psi(x)}{x},$$

to obtain $\psi(x) \sim x$, as required. $\qquad\square$

14.8   Step 7 – The Finish

Lemma 14.7. *Let $h(x)$ be either a constant or a monotone decreasing function of x. Assume $h(x)$ has the property that, for some fixed $K > 0$ and for all $x > K$,*

$$\left|\frac{\int_0^x \psi(t)\,dt - \frac{x^2}{2}}{\frac{x^2}{2}}\right| \le h(x). \tag{14.6}$$

Then, for $x > K$ and any $\epsilon > 0$

$$-\frac{\epsilon}{2} - h(x) - \frac{h(x)}{\epsilon} \le \frac{\psi(x) - x}{x} \le \frac{\epsilon}{2} + h(x) + \frac{h(x)\epsilon}{2} + \frac{h(x)}{\epsilon}.$$

Proof. Equation (14.6) is equivalent to the following

$$[1 - h(x)]\frac{x^2}{2} \le \int_0^x \psi(t)\,dt \le [1 + h(x)]\frac{x^2}{2}.$$

For our fixed $K > 0$, we now assume $y > x > K$ and define

$$F(x,y) = \int_0^y \psi(t)\,dt - \int_0^x \psi(t)\,dt = \int_x^y \psi(t)\,dt \ge 0.$$

Using properties of integrals of monotone increasing functions (such as $\psi(t)$), we have

$$(1 - h(x))\frac{y^2}{2} - (1 + h(x))\frac{x^2}{2} = (1 - h(x))\left(\frac{y^2 - x^2}{2}\right) - 2h(x)\frac{x^2}{2} \le F(x,y) \le (y - x)\psi(y)$$

$$(1 + h(x))\frac{y^2}{2} - (1 - h(x))\frac{x^2}{2} = (1 + h(x))\left(\frac{y^2 - x^2}{2}\right) + 2h(x)\frac{x^2}{2} \ge F(x,y) \ge (y - x)\psi(x).$$

The above two inequalities are easily manipulated to give

$$\psi(x) - x \le \left(\frac{y-x}{2}\right) + (h(x))\left(\frac{y+x}{2}\right) + h(x)\frac{x^2}{y-x} \tag{14.7}$$

$$\psi(y) - y \ge -\left(\frac{y-x}{2}\right) - (h(x))\left(\frac{y+x}{2}\right) - h(x)\frac{x^2}{y-x}. \tag{14.8}$$

For equation (14.7), set $y = x(1 + \epsilon)$, giving

$$\psi(x) - x \le \left(\frac{(1+\epsilon)x - x}{2}\right) + (h(x))\left(\frac{(1+\epsilon)x + x}{2}\right) + h(x)\frac{x^2}{(1+\epsilon)x - x}$$

$$\le \frac{\epsilon x}{2} + (h(x))\left(\frac{2x + \epsilon x}{2}\right) + h(x)\frac{x}{\epsilon}$$

$$\boxed{\frac{\psi(x) - x}{x}} \le \frac{\epsilon}{2} + (h(x))\left(1 + \frac{\epsilon}{2}\right) + \frac{h(x)}{\epsilon} \boxed{\le \frac{\epsilon}{2} + h(x) + \frac{h(x)\epsilon}{2} + \frac{h(x)}{\epsilon}}$$

For equation (14.8), set $x = y(1 - \epsilon)$, giving

$$\psi(y) - y \ge -\left(\frac{y - [1-\epsilon]y}{2}\right) - (h(x))\left(\frac{y + [1-\epsilon]y}{2}\right) - h(x)\frac{([1-\epsilon]y)^2}{y - [1-\epsilon]y}$$

$$\ge -\frac{y\epsilon}{2} - (h(x))\left(\frac{2y - y\epsilon}{2}\right) - h(x)\frac{(1-\epsilon)^2 \cdot y^2}{y\epsilon}$$

$$\frac{\psi(y) - y}{y} \ge -\frac{\epsilon}{2} - (h(x))\left(1 - \frac{\epsilon}{2}\right) - h(x)\frac{(1-\epsilon)^2 \cdot}{\epsilon} \ge -\frac{\epsilon}{2} - h(x) + \frac{h(x)\epsilon}{2} - \frac{h(x)}{\epsilon}$$

$$\boxed{\frac{\psi(y) - y}{y}} \boxed{\ge -\frac{\epsilon}{2} - h(x) - \frac{h(x)}{\epsilon}} \qquad \square$$

Lemma 14.8. *The following asymptotically equivalent relationship holds:* $\psi(x) \sim x$.

Proof. We must show

$$\lim_{x \to \infty} \left|\frac{\psi(x) - x}{x}\right| = 0.$$

Fix ϵ with $0 < \epsilon < 1$ and define $h(x) = \epsilon^2$. From lemma 14.5 and lemma 14.7, there is a $K > 0$ such that for all $x > K$ we have

$$-\frac{\epsilon}{2} - h(x) - \frac{h(x)}{\epsilon} \le \frac{\psi(x) - x}{x} \le \frac{\epsilon}{2} + h(x) + \frac{h(x)\epsilon}{2} + \frac{h(x)}{\epsilon}$$

Substituting for $h(x)$ gives (noting that $\epsilon^3 < \epsilon^2 < \epsilon$)

$$-\frac{\epsilon}{2} - \epsilon^2 - \frac{\epsilon^2}{\epsilon} \le \frac{\psi(x) - x}{x} \le \frac{\epsilon}{2} + \epsilon^2 + \frac{\epsilon^2\epsilon}{2} + \frac{\epsilon^2}{\epsilon}$$

$$-\frac{\epsilon}{2} - \epsilon - \epsilon \le \frac{\psi(x) - x}{x} \le \frac{\epsilon}{2} + \epsilon + \frac{\epsilon}{2} + \epsilon.$$

We can make ϵ (and therefore $h(x)$) as small as we like. Thus, we have our desired result. $\qquad \square$

Theorem 14.1 (Prime Number Theorem). $\pi(x) \sim \dfrac{x}{\log x}$.

Proof. From lemma 14.8 we have $\psi(x) \sim x$. Thus, from lemma 14.6, we have $\pi(x) \sim x/\log x$. $\qquad \square$

Chapter 15

Prime Number Theorem - A Modern Proof

15.1   Our Strategy

Following an approach due to D.J. Newman,[37] we will prove the *Prime Number Theorem* in seven steps, as outlined below.

Step 1 – Definitions   We define certain functions that may be used in the proof. We call these functions the "Prime Functions".

Step 2 – Properties   We prove certain properties of the Prime Functions.

Step 3 – Equivalent Statement   We prove that a certain statement regarding $\theta(x)$ (the first Chebyshev function) is equivalent to the *Prime Number Theorem*. That is, if you prove that equivalent statement, you have proved the *Prime Number Theorem*.

Step 4 – No Zeros   We prove that $\zeta(s) \neq 0$ for $Re(s) = 1$.

Step 5 – Newman's Analytic Theorem   We prove Newman's *Analytic Theorem*.

Step 6 – Evaluate Integral   Using the *Analytic Theorem*, we evaluate an integral.

Step 7 – The Finish   We put the pieces together and prove the *Prime Number Theorem*.

15.2   Step 1 – Definitions

We incorporate here all of the definitions included in section 12.2. In addition, we define

Definition 15.1. $\Phi(s)$ *is the **Phi function**, defined for $s \in \mathbb{C}$ by*

$$\Phi(s) = \sum_p \frac{\log p}{p^s} \qquad \textit{summed over an ordered list of primes p.}$$

15.3   Step 2 – Properties

Lemma 15.1. *For $\zeta(s)$, we have:*
(i) Let $\{s \in \mathbb{C} : Re(s) > 1\}$. $\zeta(s)$ (an infinite sum of fractions) is equivalent to an infinite product over the prime numbers, as follows:

$$\zeta(s) = \sum_{n=1}^{\infty} \frac{1}{n^s} = \prod_p \frac{1}{(1 - p^{-s})} \quad \textit{(where p ranges over the primes).}$$

(ii) The Dirichlet series for $\zeta(s)$, holomorphic on $Re(s) > 1$, can be analytically continued to the full complex plane, where it is holomorphic on $\mathbb{C}\backslash[1]$, with a simple pole at $s = 1$ with residue 1.

(iii) The function $\zeta(s) - \frac{1}{s-1}$ is holomorphic on all of $Re(s) > 0$ (i.e., no pole at $Re(s) = 1$).
(iv) The function $\zeta'(s)/\zeta(s) - \frac{1}{s-1}$ is holomorphic on $Re(s) \geq 1$ and meromorphic on $\mathbb{C}$.

Proof. We have (i) proved in Theorem 2.1, (ii) proved in Theorem 5.1, and (iii) proved in lemma 5.5.

For (iv), recall that for meromorphic function f, the poles of f'/f occur precisely at the poles (and zeros) of f. As stated above, $\zeta(s)$ is holomorphic on $\mathbb{C}\backslash\{1\}$, with a simple pole at $Re(s) = 1$ with residue 1. That means $\zeta'(s)/\zeta(s)$ is meromorphic on $\mathbb{C}$. As also stated above, $\zeta(s) - \frac{1}{s-1}$ is holomorphic on $Re(s) > 0$. Because $\zeta(s) \neq 0$ on $Re(s) \geq 1$ (by lemma 14.2), we therefore have $\zeta'(s)/\zeta(s) - \frac{1}{(s-1)}$ is holomorphic on $Re(s) \geq 1$ and meromorphic on $\mathbb{C}$. $\qquad\square$

Lemma 15.2. *Let $s \in \mathbb{C}$. For $\Phi(s)$, the Phi function, we have:*
(i) $\Phi(s)$ *is holomorphic on* $Re(s) > 1$.
(ii) $\Phi(s) - \frac{1}{s-1}$ *is meromorphic on* $Re(s) > 1/2$ *and holomorphic on* $Re(s) \geq 1$.
(iii) $[\Phi(s+1)/(s+1)] - 1/s$ *is meromorphic on* $Re(s) > -1/2$ *and holomorphic on* $Re(s) \geq 0$.

Some Preliminaries. In this section, we provide definitions that will be useful throughout this lemma. Unless stated otherwise, $x > 0$, $\delta > 0$ and $n, N, M \in \mathbb{N}$. We use $[x]^{p>}$ to mean the first prime number greater than x. We use $[[M(\delta)]]$ to mean smallest M such that $1/M < \delta$.

$\mathbf{L_1}(\delta)$. Fix $\delta > 0$ and define $\mathbf{L_1}(\delta) = \left[e^{4[[M(\delta)]]^4} \right]^{p>}$. By lemma 8.1, for $n \geq \mathbf{L_1}(\delta)$, $\log n \leq n^\delta$.

$\mathbf{L_2}(x, \delta)$. In this definition, we require $x > 1$, so that $\zeta(x)$ is convergent. Let $\zeta(x) = A_n(x) + B_{n+1}(x)$ where $A_n(x)$ represents the first n terms and $B_{n+1}(x)$ represents the remaining terms of $\zeta(x)$. Note that $B_{n+1}(x)$ is monotone decreasing as $n \to \infty$. Thus, for any $x > 1$ and for any $\delta > 0$, there must be a smallest N such that, for $n \geq N$, $B_{n+1}(x) < \delta$. Using that N, we define $\mathbf{L_2}(x, \delta) = [N]^{p>}$.

$\mathbf{L_3}(\delta)$. Fix $\delta > 0$. In this definition, we require $x > 1/2$. Define $\mathbf{L_3}(\delta) = \left[\max(2^{1/\delta}, 4) \right]^{p>}$. Then, for $n \geq \mathbf{L_3}(\delta)$, $n^{x-\delta} < n^x - 1$. To see that, consider

$$n^{x-\delta} < n^x - 1 \Longleftrightarrow n^{-x}\left(n^{x-\delta}\right) < n^{-x}\left(n^x - 1\right) \Longleftrightarrow n^{-\delta} < 1 - n^{-x} \Longleftrightarrow n^{-\delta} + n^{-x} < 1.$$

We have our result because n is large enough so that $n^{-\delta} < 1/2$ and $n^{-x} < 1/2$. $\qquad\square$

Proof (i). Fix $\lambda > 1$, set $\epsilon = \lambda - 1$ and let $Re(s) \geq \lambda > 1$. We first show that $\Phi(s)$ is absolutely convergent. For $N \in \mathbb{N}$, let $N = \mathbf{L_1}(\epsilon/2)$. We have

$$\sum_p \left| \frac{\log p}{p^s} \right| \leq \sum_{n=2}^{\infty} \frac{\log n}{n^{Re(s)}} \leq \sum_{n=2}^{N} \frac{\log n}{n^{1+\epsilon}} + \sum_{n=N+1}^{\infty} \left[\frac{\log n}{n^{\epsilon/2}} \cdot \frac{1}{n^{1+\epsilon/2}} \right] \leq \sum_{n=2}^{N} \frac{\log n}{n^{1+\epsilon}} + \sum_{n=N+1}^{\infty} \frac{1}{n^{1+\epsilon/2}}.$$

In the final inequality, we can disregard the first N terms. We have our convergence by comparison to $\zeta(1 + \epsilon/2)$. Since we have absolute convergence for any $\lambda > 1$, $\Phi(s)$ is absolutely convergent on $Re(s) > 1$.

We next show that $\Phi(s)$ is uniformly convergent. Let $N = \mathbf{L_1}(\epsilon/2)$. We will use the Weierstrass M-test, setting $M_n = n^{-(1+\epsilon)} \log n$, so that

$$\sum_{n=1}^{\infty} M_n = \sum_{n=1}^{\infty} \frac{\log n}{n^{(1+\epsilon)}} = \sum_{n=1}^{N} \frac{\log n}{n^{(1+\epsilon)}} + \sum_{n=N+1}^{\infty} \frac{\log n}{n^{(1+\epsilon)}} \leq \sum_{n=1}^{N} \frac{\log n}{n^{(1+\epsilon)}} + \sum_{n=N+1}^{\infty} \frac{1}{n^{(1+\epsilon/2)}}.$$

Again, we can disregard the first N terms and have our convergence by comparison to $\zeta(1+\epsilon/2)$. Thus, by the Weierstrass M-test, $\Phi(s)$ is uniformly convergent for all $Re(s) \geq \lambda > 1$. Since we have uniform convergence for any $\lambda > 1$, $\Phi(s)$ converges uniformly on $Re(s) > 1$.

With absolute and uniform convergence established, we are ready to show that $\Phi(s)$ is holomorphic on $Re(s) > 1$. Define

$$\Phi_n(s) = \sum_{k=1}^{n} \frac{\log p_k}{p_k^s} \qquad \text{for } Re(s) > 1 \text{, where } p_k \text{ is the } k\text{th prime in an ordered list of primes,}$$

and note that $\Phi_n(s)$ is a finite sum of holomorphic functions and therefore holomorphic. Thus, if we can establish that $\lim_{n\to\infty} \Phi_n(s) \to \Phi(s)$ for $Re(s) > 1$, then $\Phi(s)$ is holomorphic on $Re(s) > 1$.

Fix $\delta > 0$ and set $M = \max(\mathbf{L_1}(\epsilon/2), \mathbf{L_2}(1+\epsilon/2, \delta))$. Assume the Nth prime $p_N = [M]^{p>}$. We have

$$|\Phi_N(s) - \Phi(s)| \leq \sum_{n=N+1}^{\infty} \frac{\log n}{n^{Re(s)}} \leq \sum_{n=N+1}^{\infty} \frac{\log n}{n^{1+\epsilon}} = \sum_{n=N+1}^{\infty} \left[\frac{\log n}{n^{\epsilon/2}} \cdot \frac{1}{n^{1+\epsilon/2}} \right] \leq \sum_{n=N+1}^{\infty} \frac{1}{n^{1+\epsilon/2}} < \delta,$$

as required. We can conclude that $\Phi(s)$ is holomorphic on $Re(s) > 1$. $\qquad\square$

Proof (ii). We start by proving that the following infinite series is holomorphic on $Re(s) > 1/2$:

$$P(s) = \sum_p \frac{\log p}{p^s(p^s - 1)} \qquad \text{where } p \text{ is an ordered list of all primes.} \tag{15.1}$$

Fix $\lambda > 1/2$, set $\epsilon = \lambda - 1/2$ and let $x = Re(s) \geq \lambda > 1/2$. Noting that $|a - b| \geq |a| - |b|$, we have

$$|P(s)| = \sum_p \left| \frac{\log p}{p^s(p^s - 1)} \right| \leq \sum_{n=2}^{\infty} \frac{\log n}{|n^s| |(n^s - 1)|} \leq \sum_{n=2}^{\infty} \frac{\log n}{n^x (|n^s| - |1|)} = \sum_{n=2}^{\infty} \frac{\log n}{n^x (n^x - 1)}.$$

Now set $N = \max(\mathbf{L_1}(\epsilon/2), \mathbf{L_3}(\epsilon/2))$, and call the first N terms A_N

$$\leq \sum_{n=2}^{\infty} \left(\frac{\log n}{n^{1/2+\epsilon}} \cdot \frac{1}{\left(n^{1/2+\epsilon} - 1\right)} \right) \leq A_N + \sum_{n=N+1}^{\infty} \left(\frac{1}{n^{1/2+\epsilon/2}} \cdot \frac{1}{n^{1/2+\epsilon/2}} \right) = A_N + \sum_{n=N+1}^{\infty} \frac{1}{n^{1+\epsilon}}.$$

Thus, $P(s)$ is absolutely convergent by comparison to the absolutely convergent $\zeta(1 + \epsilon)$. By similar reasoning, we use the Weierstrass M-test, setting $M_n = \log n / n^\lambda (n^\lambda - 1)$, to show that that $P(s)$ is uniformly convergent.

To see that $P(s)$ is holomorphic on $Re(s) > 1/2$, define $P_n(s)$ as the first n terms of $P(s)$. Fix $\delta > 0$ and set $N = \max(\mathbf{L_1}(\epsilon/2), \mathbf{L_2}(1+\epsilon/2, \delta))$. We then have that $|P_n(s) - P(s)| < \delta$ for $n > N$, as required.

With $P(s)$ shown holomorphic on $Re(s) > 1/2$, we can complete our proof. Unless otherwise stated, we now assume $Re(s) > 1$. We have (for $p = primes$)

$$\zeta(s) = \prod_p \frac{1}{1 - p^{-s}} \qquad \text{so that} \qquad \log \zeta(s) = - \sum_p \log(1 - p^{-s}).$$

Now take the derivative

$$\frac{\zeta'(s)}{\zeta(s)} = - \sum_p \frac{\log p \cdot p^{-s}}{1 - p^{-s}} = - \sum_p \log p \left(\frac{p^{-s}}{1 - p^{-s}} \right)$$

$$-\frac{\zeta'(s)}{\zeta(s)} = \sum_p \log p \left(\frac{p^{-s}}{1 - p^{-s}} \right) = \sum_p \log p \left(\frac{1}{p^s - 1} \right).$$

Using the equality

$$\frac{1}{p^s - 1} = \frac{p^s}{p^s(p^s - 1)} = \frac{(p^s - 1) + 1}{p^s(p^s - 1)} = \frac{(p^s - 1)}{p^s(p^s - 1)} + \frac{1}{p^s(p^s - 1)} = \frac{1}{p^s} + \frac{1}{p^s(p^s - 1)},$$

we have

$$\frac{\zeta'(s)}{\zeta(s)} = \sum_p \log p \left(\frac{1}{p^s - 1} \right) = \sum_p \frac{\log p}{p^s} + \sum_p \frac{\log p}{p^s(p^s - 1)}$$

$$= \Phi(s) + \sum_p \frac{\log p}{p^s(p^s - 1)}.$$

With some rearrangement, this gives

$$\Phi(s) - \frac{1}{s - 1} = \left[\frac{\zeta'(s)}{\zeta(s)} - \frac{1}{s - 1} \right] - \sum_p \frac{\log p}{p^s(p^s - 1)}. \tag{15.2}$$

By lemma 15.1(iv), $\zeta'(s)/\zeta(s) - 1/(s-1)$ is holomorphic on $Re(s) \geq 1$ and meromorphic on $\mathbb{C}$. The remaining term on the RHS is $P(s)$, holomorphic on $Re(s) \geq 1/2$. Thus, the RHS (and therefore the LHS) is holomorphic on $Re(s) \geq 1$ and meromorphic on $Re(s) > 1/2$, as required. $\qquad\square$

Proof (iii). Define

$$G(s) = \Phi(s) - \frac{1}{s - 1} \quad \text{and} \quad H(s) = \frac{1}{s}(G(s) - 1)$$

so that

$$H(s) = \frac{1}{s}\left(\Phi(s) - \frac{1}{s - 1} - 1 \right) = \frac{1}{s}\left(\Phi(s) - \frac{s}{s - 1} \right) = \frac{\Phi(s)}{s} - \frac{1}{s - 1},$$

By lemma 15.2(ii), we have $G(s)$ [and therefore $H(s)$] is holomorphic on $Re(s) \geq 1$ and meromorphic on $Re(s) > 1/2$. Using the final equation for $H(s)$, above, we can then define

$$H^*(s) = H(s + 1) = \frac{\Phi(s + 1)}{s + 1} - \frac{1}{s},$$

and see that $H^*(s)$ is holomorphic on $Re(s) \geq 0$ and meromorphic on $Re(s) > -1/2$. $\qquad\square$

Lemma 15.3. *Let $x \in \mathbb{R}_{>0}$. For $\theta(x)$, the first Chebyshev function, $\theta(x) = \mathcal{O}(x)$.*

Proof. We must show that for some $x_0 > 0$ there is some $M > 0$ such that for $x > x_0$ we have $\theta(x) \leq Mx$.

Let $n \in \mathbb{N}$. To set up this proof, we note that

$$e^{\theta(2n) - \theta(n)} = \frac{e^{\sum_{p \leq 2n} \log p}}{e^{\sum_{p \leq n} \log p}} = \frac{\prod_{p \leq 2n} e^{\log p}}{\prod_{p \leq n} e^{\log p}} = \prod_{n < p \leq 2n} e^{\log p} = \prod_{n < p \leq 2n} p,$$

and also that

$$2^{2n} = (1 + 1)^{2n} = \binom{2n}{0} + \binom{2n}{1} + \ldots + \binom{2n}{2n} \geq \binom{2n}{n}.$$

Now observe that

$$\binom{2n}{n} = \frac{(2n)!}{n!n!} = \frac{(n+1)(n+2)\cdots(2n)}{n!} \geq \prod_{n<p\leq 2n} p.$$

We justify the above because $\binom{2n}{n}$ is an integer and all p in our product divide the numerator, while no p in our product can be factored out by the denominator. That is, the whole number $\binom{2n}{n}$ includes all p in our product as factors.

Putting all of the above observations together

$$2^{2n} \geq \prod_{n<p\leq 2n} p = e^{\theta(2n)-\theta(n)}.$$

Taking the logarithms gives

$$2n \log 2 \geq \theta(2n) - \theta(n).$$

Now using telescoping, we have

$$\theta(2^m) = \sum_{n=1}^{m} \left[\theta(2^n) - \theta(2^{n-1})\right] \leq \sum_{n=1}^{m} 2^n \log 2 = (2^{m+1} - 2) \log 2 < 2^{m+1} \log 2.$$

Now let $x \geq 1$ and choose m such that $2^{m-1} \leq x < 2^m$, so that

$$\theta(x) \leq \theta(2m) \leq 2^{m+1} \log 2 = (4 \log 2)2^{m-1} \leq (4 \log 2)x.$$

Thus, $\theta(x) = \mathcal{O}(x)$. $\qquad\square$

15.4   Step 3 – Equivalent Statement

Lemma 15.4. *If $\theta(x) \sim x$, then $\pi(x) \sim x/\log x$, and the **Prime Number Theorem** is proved.*

Proof. By the definition of $\theta(x)$, for $x \geq 1$

$$0 \leq \theta(x) \leq \pi(x) \log x,$$

and therefore

$$\frac{\theta(x)}{x} \leq \pi(x)\frac{\log x}{x}.$$

Now let $0 < \epsilon < 1$ so that

$$\theta(x) \geq \sum_{x^{1-\epsilon}<p\leq x} \log p \geq \sum_{x^{1-\epsilon}<p\leq x} \log x^{1-\epsilon} = (1-\epsilon) \log x \sum_{x^{1-\epsilon}<p\leq x} 1,$$

where we have excluded in the sum any primes below $x^{1-\epsilon}$, and then used $\log x^{1-\epsilon} \leq \log p$:

$$= (1-\epsilon) \log x \left[\pi(x) - \pi(x^{1-\epsilon})\right]$$
$$\geq (1-\epsilon) \log x \left[\pi(x) - x^{1-\epsilon}\right].$$

With some simple rearrangements we have

$$\pi(x) \le \frac{\theta(x)}{(1-\epsilon)(\log x)} + x^{1-\epsilon}.$$

Putting our results together gives

$$\frac{\theta(x)}{x} \le \pi(x)\frac{\log x}{x} \le \frac{\theta(x)}{x(1-\epsilon)} + \frac{\log x}{x^\epsilon}.$$

Applying lemma 8.1, for any $\epsilon > 0$ we have $\log x / x^\epsilon \to 0$ as $x \to \infty$. Therefore, it follows from our last equation that $\theta(x)/x \to 1$ if and only if $\pi(x) \cdot \log x / x \to 1$, as required. $\qquad\square$

15.5   Step 4 – No Zeros

The proof here is identical to the proof in lemma 14.2, which was used in the "Classic" proof of the *Prime Number Theorem*.

15.6   Step 5 – Newman's Analytic Theorem

Lemma 15.5. *Fix $K > 0$, let $s \in \mathbb{C}$ and assume $f(s)$ is holomorphic on $Re(s) \ge 0$. Then there exists an $\epsilon > 0$ such that $f(s)$ is holomorphic on $\{s \in \mathbb{C} : |Im(s)| \le K, Re(s) \ge -\epsilon\}$.*

Proof. Define the line segment $L = \{s \in \mathbb{C} : |Im(s)| \le K, Re(s) = 0\}$. Because $f(s)$ is holomorphic at each point on L, it is also holomorphic in a neighborhood of each point on L. Thus, for each $s \in L$, there must be a $r_s > 0$ such that $f(s)$ is holomorphic in the open disk $B(r_s, s) = \{z \in \mathbb{C} : |z - s| < r_s\}$. We therefore have

$$L \subset \bigcup_{s \in L} B(r_s, s)$$

where our union of open disks is an open cover of L. But L is closed and bounded and therefore compact. Thus, by Heine–Borel, a finite subcover of L can be constructed from the above open cover. In particular, for some $n \in \mathbb{N}$, there exists a set $S = s_1, s_2, ... s_n \in L$, ordered by increasing $Im(s)$, with

$$L \subset \bigcup_{k=1}^{n} B(r_{s_k}, s_k) \quad \text{for } s_k \in S.$$

For convenience in constructing our finite subcover, we modify our set S slightly as follows: (1) if not already included, we assume our set S includes the endpoints $-iK$ and iK, (2) we assume that, for any two points s_p and s_q in S, we have discarded any such s_p where $B(r_{s_p}, s_p) \subseteq B(r_{s_q}, s_q)$, but only if s_p is not an endpoint, and (3) we assume our set S consists of m unique points in L.

For any two adjacent points s_j and s_{j+1} in our set S, consider the line segment

$$L_j = \{s \in \mathbb{C} : Re(s) = 0, Im(s_j) \le s \le Im(s_{j+1})\} \quad \text{where } L_j \subseteq L.$$

Now construct a point p in L_j as follows. Let $d_j = Im(s_{j+1}) - Im(s_j)$. Let $p_1 = s_j + i\min(r_{s_j}, d_j)$. Let $p_2 = s_{j+1} - i\min(r_{s_{j+1}}, d_j)$. Finally, let $p = (p_1 + p_2)/2$ so that $p \in B(r_{s_j}, s_j) \cap B(r_{s_{j+1}}, s_{j+1})$. Note that the intersection of two open sets is also open. That means there is an $r_p > 0$ such that

$$B(r_p, p) \subset B(r_{s_j}, s_j) \cap B(r_{s_{j+1}}, s_{j+1}).$$

Next define the closed rectangle $R_j = \{s \in \mathbb{C} : -\epsilon_j \leq Re(s) \leq 0, Im(s_j) \leq s \leq Im(s_{j+1})\}$, where $\epsilon_j = r_p/2$. Because $Im(p) > Im(s_j)$ and $Im(p) < Im(s_{j+1})$, we can use the convexity of our disks to conclude that $R_j \subset B(r_{s_j}, s_j) \cup B(r_{s_{j+1}}, s_{j+1})$.

Our line segment L_j was arbitrarily chosen, so we can similarly construct a rectangle like R_j for each of the $m - 1$ line segments that make up L. Let $\epsilon = \min(\epsilon_1, \epsilon_2, ... \epsilon_{m-1})$. If we define

$$R = \{s \in \mathbb{C} : -\epsilon \leq Re(s) \leq 0, -K \leq Im(s) \leq K\},$$

then we have

$$R \subset \bigcup_{k=1}^{m} B(r_{s_k}, s_k) \quad \text{for } s_k \in S.$$

We can conclude that $f(s)$ is holomorphic on R, as required. $\qquad \square$

Theorem 15.1 (Newman's Analytic Theorem). *Let $t \in \mathbb{R}$ and $z \in \mathbb{C}$. For $t \geq 0$, let $f(t)$ be a bounded and piecewise continuous function, and suppose that, for $Re(z) > 0$, $g(z) = \int_0^\infty f(t)e^{-zt}\,dt$ extends holomorphically to $Re(z) \geq 0$. Then the improper integral $\int_0^\infty f(t)\,dt$ converges and equals $g(0)$.*

Proof. We start by defining

$$g_T(z) = \int_0^T f(t)e^{-zt}\,dt \quad \text{for } 0 < T < \infty.$$

It is easily seen that $g_T(z)$ is entire for any given T. Our goal is to show that

$$g_T(0) = \int_0^T f(t)e^0\,dt = \int_0^T f(t)\,dt \to g(0) \text{ as } T \to \infty.$$

Next we define the contour of integration C and the helper function $H(z)$ as follow.

Fix (large) $R > 1$. From Lemma 15.5, there is an $\epsilon > 0$ such that $g(z)$ is holomorphic inside and on the rectangle

$$Rect_R = \{z \in \mathbb{C} : -\epsilon \leq Re(z) \leq 0, -R \leq Im(z) \leq R\}.$$

Define the region

$$D_R = \{z \in \mathbb{C} : |z| < R, Re(z) > -\epsilon\}.$$

Clearly, $g(z)$ is holomorphic inside and on the boundary of D_R. Now let C be a counterclockwise oriented closed path which is the boundary of D_R, and define

$$C^+ = C \cap \{Re(z) \geq 0\} \quad \text{and} \quad C^- = C \cap \{Re(z) < 0\}.$$

For our helper function, we define

$$H_T(z) = \left(g(z) - g_T(z)\right) e^{zT} \left(1 + \frac{z^2}{R^2}\right).$$

The natural choice of $H_T(z) = g(z) - g_T(z)$ runs into bounding problems. It will become clear below that Newman's clever choice of $H_T(z)$ is needed for our proof. Note that $H_T(z)$ is holomorphic in and on C, and $H_T(0) = g(0) - g_T(0)$. Applying the Cauchy integral formula, we have

$$H_T(0) = g(0) - g_T(0) = \frac{1}{2\pi i} \int_C \frac{H_T(z)}{z - 0}\,dz = \frac{1}{2\pi i} \int_C \left(g(z) - g_T(z)\right) e^{zT} \left(1 + \frac{z^2}{R^2}\right) \frac{dz}{z}. \quad (15.3)$$

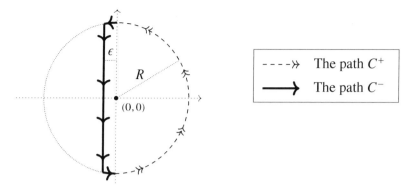

Figure 15.1: Contour for Analytic Theorem

We first consider the integral over C^+ where we are on the semicircle with $|z| = R$. Note that on the circle, we have $R^2 = |z|^2 = z \cdot \bar{z}$, allowing the following simplification

$$\left| \left(1 + \frac{z^2}{R^2} \right) \frac{1}{z} \right| = \left| \frac{R^2 + z^2}{R^2 z} \right| = \left| \frac{|z|^2 + z^2}{R^2 z} \right| = \left| \frac{z(\bar{z} + z)}{R^2 z} \right| = \left| \frac{\bar{z} + z}{R^2} \right| = \frac{2|Re(z)|}{R^2},$$

so that

$$\left| e^{zT} \left(1 + \frac{z^2}{R^2} \right) \frac{1}{z} \right| = e^{Re(z)T} \cdot \frac{2Re(z)}{R^2}. \tag{15.4}$$

Now define $B = max_{t \geq 0}|f(t)|$, giving

$$|g(z) - g_T(z)| = \left| \int_T^\infty f(t)e^{-zt}\, dt \right| \leq B \int_T^\infty |e^{-zt}|\, dt = \frac{Be^{-Re(z)T}}{Re(z)} \quad Re(z) > 0. \tag{15.5}$$

Combining the above, we see that $|H_T(z)/z| < 2B/R^2$ for $z \in C^+$, providing a bound for the absolute value of the integrand in equation (15.3) on C^+. By continuity, the result is valid for $Re(z) \geq 0$ because $g(z) - g_T(z)$ is analytic on $Re(z) \geq 0$; also the $Re(z)$ in the denominator of equation (15.5) is canceled by equation (15.4). And that means (by the ML Inequality)

$$\left| \frac{1}{2\pi i} \int_{C^+} (g(z) - g_T(z))\, e^{zT} \left(1 + \frac{z^2}{R^2} \right) \frac{dz}{z} \right| \leq \frac{B}{R}. \tag{15.6}$$

Note that the estimate of the integral along the path C^+ is independent of T.

We now consider the integral over C^- by looking at $g(z)$ and $g_T(z)$ separately. We start with an estimate for $|g_T(z)|$:

$$|g_T(z)| = \left| \int_0^T f(t)e^{-zT}\, dt \right| \leq B \int_0^T |e^{-zT}|\, dt = \frac{Be^{-Re(z)T}}{Re(z)} \quad Re(z) < 0.$$

Since $g_T(z)$ is entire, the path of integration C^- can be (temporarily) replaced by the semicircle

$$SC^- = \{z \in \mathbb{C} : |z| = R, Re(z) < 0\},$$

where we again obtain equation (15.4). Combining the two results, we have for $g_T(z)$ on C^-

$$\left| \frac{1}{2\pi i} \int_{C^-} g_T(z)e^{zT} \left(1 + \frac{z^2}{R^2} \right) \frac{dz}{z} \right| \leq \frac{B}{R}. \tag{15.7}$$

Finally, we need to estimate $g(z)$ on C^-

$$\left| \frac{1}{2\pi i} \int_{C^-} g(z)e^{zT} \left(1 + \frac{z^2}{R^2} \right) \frac{dz}{z} \right|.$$

We have that $g(z)$ is analytic (and therefore bounded) on C^-. Thus, there is some constant K such that

$$\left| g(z) \left(1 + \frac{z^2}{R^2} \right) \frac{1}{z} \right| = \left| g(z) \frac{2|Re(z)|}{R^2} \right| \le K \quad \text{for } z \in C^-,$$

where again that the bound is independent of T.

The remaining term, $\left| e^{zT} \right|$, goes to zero uniformly on any compact subset of C^- as $T \to \infty$, because on any such compact subset $Re(z) < 0 \Rightarrow 0 < |e^z| < x_1 < 1$ for some bound x_1. Therefore

$$\lim_{T \to \infty} \frac{1}{2\pi i} \int_{C^-} g(z)e^{zT} \left(1 + \frac{z^2}{R^2} \right) \frac{dz}{z} = 0,$$

from which it follows that

$$\limsup_{T \to \infty} |g(0) - g_T(0)| \le \frac{2B}{R}.$$

Now recall that B is fixed and independent of R and T. The selected R (also independent of T) was arbitrary. Letting $R \to \infty$, we have our desired result

$$\lim_{T \to \infty} g_T(0) = g(0). \qquad \square$$

15.7   Step 6 – Evaluate Integral

Lemma 15.6. *The following integral converges* $\int_1^\infty \frac{\theta(x) - x}{x^2} \, dx$.

Proof. We start with a change of variable $x = e^t$ ($t = \log x$) so that $dt = dx/x$:

$$\int_1^\infty \frac{\theta(x) - x}{x^2} \, dx = \int_1^\infty \left(\frac{\theta(x)}{x} - 1 \right) \frac{dx}{x} = \int_0^\infty \left(\theta(e^t)e^{-t} - 1 \right) dt = \int_0^\infty f(t) \, dt,$$

where $f(t) = \theta(e^t)e^{-t} - 1$.

Now define $g(z) = \int_0^\infty f(t)e^{-zt} \, dt$ for all $Re(z) \ge 0$. We have set up functions $f(t)$ and $g(z)$ to use the *Analytic Theorem* (Theorem 15.1). It remains to show that we have met the required conditions of that theorem.

We start with $f(t)$. We must show that it is bounded and locally integrable. From lemma 15.3, there is some constant $M > 0$ such that

$$|f(t)| \le |\theta(e^t)|e^{-t} + 1 \le (Me^t)e^{-t} + 1 = M + 1,$$

for all t greater than some t_0. Since $f(t)$ has only jump discontinuities when t is a prime, it is clearly bounded on $(0, t_0]$ and therefore bounded on $(0, \infty)$. Also, for any compact $K \subset (0, \infty)$, $f(t)$ has only finitely many discontinuities, so the integral of $f(t)$ over K is finite. Therefore $f(t)$ is locally integrable and $f(t)$ satisfies the conditions of the *Analytic Theorem*.

It remains to show that $g(z)$ is holomorphic on $Re(z) \geq 0$. To do so, consider the related integral below, where we let J be the value of the integral. Unless otherwise indicated, we assume $Re(z) > 0$.

$$J = \int_0^\infty \theta(e^t) e^{-(z+1)t} \, dt.$$

Set $s = z + 1$ for $Re(s) > 1$ and make the change of variable $x = e^t$ ($t = \log x$) so that $dt = dx/x$.

$$J = \int_0^\infty \theta(e^t) e^{-(z+1)t} \, dt = \int_0^\infty \theta(e^t) \cdot (e^t)^{-(z+1)} \, dt = \int_1^\infty \theta(x) x^{-(z+1)} \frac{dx}{x} = \int_1^\infty \frac{\theta(x)}{x^{s+1}} \, dx.$$

Rewrite the last equality, using $p_1, p_2, \ldots$ as an enumeration of the primes and $p_0 = 1$. We also add a factor of s on both sides

$$sJ = \int_1^\infty \frac{s\theta(x)}{x^{s+1}} \, dx = \sum_{k=0}^\infty \int_{p_k}^{p_{k+1}} \frac{s\theta(x)}{x^{s+1}} \, dx.$$

Next, note that for $x \in (p_k, p_{k+1})$ we have

$$\theta(x) = \sum_{p \leq x} \log p = \sum_{p \leq p_k} \log p = \theta(p_k),$$

allowing us to shift $\theta(x)$ outside of the integral:

$$sJ = \sum_{k=0}^\infty \theta(p_k) \int_{p_k}^{p_{k+1}} \frac{s}{x^{s+1}} \, dx = \sum_{k=0}^\infty \theta(p_k) \left(\frac{1}{p_k^s} - \frac{1}{p_{k+1}^s} \right).$$

Our next refinement is to note that $\theta(p_k) = \theta(p_{k+1}) - \log(p_{k+1})$, allowing

$$sJ = \sum_{k=0}^\infty \left(\frac{\theta(p_k)}{p_k^s} - \frac{\theta(p_{k+1})}{p_{k+1}^s} + \frac{\log(p_{k+1})}{p_{k+1}^s} \right) = \sum_{k=0}^\infty \left(\frac{\theta(p_k)}{p_k^s} - \frac{\theta(p_{k+1})}{p_{k+1}^s} \right) + \sum_{k=1}^\infty \frac{\log(p_k)}{p_k^s}.$$

To justify dividing the sum into two sums, we must show that both sums are convergent. The second sum is equal to $\Phi(s)$, shown convergent in lemma 15.2.

For the first sum, we will use telescoping, which proves the sum is convergent if the resulting limit (below) is convergent. We also use lemma 15.3, which states that for all large enough k there is an $M > 0$ such that $\theta(p_k) < M \cdot p_k$. Thus, for $Re(s) > 1$

$$\sum_{k=0}^\infty \left(\frac{\theta(p_k)}{p_k^s} - \frac{\theta(p_{k+1})}{p_{k+1}^s} \right) = \frac{\theta(p_0)}{p_0^s} + \lim_{k \to \infty} \frac{\theta(p_k)}{p_k^s} = \frac{0}{1} + \lim_{k \to \infty} \frac{\theta(p_k)}{p_k^s} \leq \lim_{k \to \infty} \frac{M \cdot p_k}{p_k^s} = \lim_{k \to \infty} \frac{M}{p_k^{s-1}} = 0.$$

For $Re(s) > 1$, we have shown that $sJ = \Phi(s)$. We reverse our change of variable back to $s = z + 1$. Therefore, for $Re(z) > 0$, we have $J = \Phi(z+1)/(z+1)$.

We can now return to $g(z)$. We have

$$g(z) = \int_0^\infty f(t) e^{-zt} \, dt = \int_0^\infty (\theta(e^t) e^{-t} - 1) e^{-zt} \, dt = \int_0^\infty \theta(e^t) e^{-t} e^{-zt} - e^{-zt} \, dt$$

$$= \int_0^\infty \theta(e^t) e^{-(z+1)t} - e^{-zt} \, dt = \int_0^\infty \theta(e^t) e^{-(z+1)t} \, dt - \int_0^\infty e^{-zt} \, dt = \frac{\Phi(z+1)}{(z+1)} - \int_0^\infty e^{-zt} \, dt$$

$$= \frac{\Phi(z+1)}{(z+1)} - \frac{1}{z}.$$

With our last equation, we apply lemma 15.2(iii) to show that $g(z)$ is holomorphic on $Re(z) \geq 0$. $\square$

15.8   Step 7 – The Finish

We have assembled all of the pieces needed to prove the *Prime Number Theorem*.

Theorem 15.2 (Prime Number Theorem). $\pi(x) \sim \dfrac{x}{\log x}$.

Proof. By lemma 15.4, we need only prove $\theta(x) \sim x$; that is, for any small $\epsilon > 0$, there is an M such that for all $x > M$:

$$1 - \epsilon < \theta(x)/x < 1 + \epsilon.$$

Our proof will be by contradiction. Fix ϵ and assume that $\theta(x)$ is **not** asymptotic to x. That means one or both of the following are true: (1) there exists an arbitrarily large x such that $\theta(x) \leq (1 - \epsilon)x$, and/or (2) there exists an arbitrarily large x such that $\theta(x) \geq (1 + \epsilon)x$. We consider the two cases separately.

Case 1. Fix ϵ and let x_1 be our arbitrarily large x such that $\theta(x_1) \leq (1 - \epsilon)x_1$. At the outset, it is important to note that x_1 cannot be the largest x satisfying the $\theta(x) \leq (1 - \epsilon)x$ hypothesis. Otherwise, we could just set M to $x_1 + 1$ and have our asymptotic condition. Thus, there must be an $x_2 > x_1$, an $x_3 > x_2$ and in fact an infinite number of $x_n's$, with no limit on their value. We can therefore assume that our x_1 is as large as necessary.

Because $\theta(x)$ is non-decreasing, we have

$$\int_{(1-\epsilon)x_1}^{x_1} \frac{\theta(u) - u}{u^2}\, du \leq \int_{(1-\epsilon)x_1}^{x_1} \frac{\theta(x_1) - u}{u^2}\, du \leq \int_{(1-\epsilon)x_1}^{x_1} \frac{(1-\epsilon)x_1 - u}{u^2}\, du$$

$$= (1-\epsilon)x_1 \int_{(1-\epsilon)x_1}^{x_1} \frac{1}{u^2}\, du - \int_{(1-\epsilon)x_1}^{x_1} \frac{1}{u}\, du = \left[\frac{-(1-\epsilon)x_1}{u} - \log u \right]_{(1-\epsilon)x_1}^{x_1}$$

$$= -(1-\epsilon) + 1 - \log x_1 + \log(1-\epsilon)x_1 = \epsilon + \log(1-\epsilon).$$

For $0 < (1 - \epsilon) < 1$, the Taylor Series for $\log(1 - \epsilon)$ shows that there is a constant $0 < C_\epsilon < 1$ such that $\log(1 - \epsilon) = -\epsilon - C_\epsilon$. Therefore, our integral is negative and equal to $-C_\epsilon$.

On the other hand, by lemma 15.6, the integral below converges for any $1 \leq A < B$. And that means, under the Cauchy criterion, that for every $\delta > 0$ there is an $N > 1$ such that for all $A, B \geq N$ we have

$$\left| \int_A^B \frac{\theta(u) - u}{u^2}\, du \right| < \delta.$$

Let $\delta = C_\epsilon/2$ to obtain the necessary N. With x_1 arbitrarily large, we can then set $A = (1 - \epsilon)x_1$ and $B = x_1$ and we have our contradiction.

Case 2. Fix ϵ and let x_1 be our arbitrarily large x such that $\theta(x_1) \geq (1 + \epsilon)x_1$. As discussed above, we can assume that our x_1 is as large as necessary.

Because $\theta(x)$ is non-decreasing, we have

$$\int_{x_1}^{(1+\epsilon)x_1} \frac{\theta(u) - u}{u^2}\, du \geq \int_{x_1}^{(1+\epsilon)x_1} \frac{\theta(x_1) - u}{u^2}\, du \geq \int_{x_1}^{(1+\epsilon)x_1} \frac{(1+\epsilon)x_1 - u}{u^2}\, du$$

$$= (1+\epsilon)x_1 \int_{x_1}^{(1+\epsilon)x_1} \frac{1}{u^2}\, du - \int_{x_1}^{(1+\epsilon)x_1} \frac{1}{u}\, du = \left[\frac{-(1+\epsilon)x_1}{u} - \log u \right]_{x_1}^{(1+\epsilon)x_1}$$

$$= -1 + (1+\epsilon) + \log x_1 - \log(1+\epsilon)x_1 = \epsilon - \log(1+\epsilon)$$

For $0 < \epsilon < 1$, the Taylor Series for $\log(1 + \epsilon)$ shows that there is a constant $0 < C_\epsilon < 1$ such that $\log(1 + \epsilon) = \epsilon - C_\epsilon$, thus ensuring that our integral is positive and equal to C_ϵ.

We now apply the Cauchy criterion logic of Case 1 to obtain our contradiction. Case 2 cannot be true. Our proof by contradiction is complete, so we must have $\theta(x) \sim x$, thereby proving the theorem. $\square$

Chapter 16

After the Prime Number Theorem

After the *Prime Number Theorem* was proved, attention turned to finding a zero-free region *inside* the strip $0 \leq Re(s) \leq 1$. That is, for $s = \beta + i\gamma$, finding a positive, real-valued function $f(\gamma)$ such that $\zeta(s) \neq 0$ for $\beta \leq 1 - f(\gamma)$. There have been many improvements on $f(\gamma)$ over the years. In section 16.1, we present a variation on the very first proof of a zero-free region.

There is a close relationship between the zero-free region, above, and the *error term* of the asymptotic functions associated with the *Prime Number Theorem*. In section 16.2, we present a variation on the first proof showing an error term for $\psi(x) \sim x$ by use of the zero-free region.

Finally, with proof of our simple error term completed, in section 16.3, we show that the Riemann Hypothesis is equivalent to a statement about the error term of $|\psi(x) \sim x|$.

16.1 Zero-Free Region for $\zeta(s)$

We have seen in Chapters 14 and 15 that, to prove the *Prime Number Theorem*, you must show that $\zeta(s) \neq 0$ for $Re(s) = 1$. An improvement is to show a zero-free region for $\zeta(s)$ *inside* the "critical strip" $0 \leq Re(s) \leq 1$. We present here a proof that shows the existence of such a zero-free region[38].

16.1.1 Some Useful Lemmas

Lemma 16.1. *Let $\{s = \sigma + it \in \mathbb{C} : \sigma > 1\}$. Then*

$$Re\left[3\frac{\zeta'(\sigma)}{\zeta(\sigma)} + 4\frac{\zeta'(\sigma + it)}{\zeta(\sigma + it)} + \frac{\zeta'(\sigma + 2it)}{\zeta(\sigma + 2it)}\right] \leq 0.$$

Proof. We start with the formula from lemma 12.1:

$$-\frac{\zeta'(s)}{\zeta(s)} = s \int_1^\infty \psi(x) x^{-s-1} \, dx,$$

and use $x^s = x^{\sigma+it} = e^{(\sigma+it)\log x} = e^{\sigma \log x} e^{it \log x} = x^\sigma \cdot (cos(t\log x) + i\sin(t\log x))$. Therefore $Re(x^s) = x^\sigma \cdot cos(t\log x)$. That allows

$$Re\left[-3\frac{\zeta'(\sigma)}{\zeta(\sigma)} - 4\frac{\zeta'(\sigma + it)}{\zeta(\sigma + it)} - \frac{\zeta'(\sigma + 2it)}{\zeta(\sigma + 2it)}\right] = \int_1^\infty \psi(x) x^{-\sigma-1} \left[3 + 4\cos(t\log x) + \cos(2t\log x)\right] \, dx.$$

Using lemma 14.1, we have that $M(x) = [3 + 4\cos(t\log x) + \cos(2t\log x)] \geq 0$. Thus

$$Re\left[-3\frac{\zeta'(\sigma)}{\zeta(\sigma)} - 4\frac{\zeta'(\sigma + it)}{\zeta(\sigma + it)} - \frac{\zeta'(\sigma + 2it)}{\zeta(\sigma + 2it)}\right] = \int_1^\infty \psi(x) x^{-\sigma-1} [M(x)] \, dx \geq 0,$$

giving our result (for $\sigma > 1$)

$$Re\left[3\frac{\zeta'(\sigma)}{\zeta(\sigma)} + 4\frac{\zeta'(\sigma + it)}{\zeta(\sigma + it)} + \frac{\zeta'(\sigma + 2it)}{\zeta(\sigma + 2it)}\right] \le 0.$$ $\square$

Lemma 16.2. *Let* $\sigma > 1$. *Then* $-\dfrac{\zeta'(\sigma)}{\zeta(\sigma)} \le \dfrac{1}{\sigma - 1}$ *(or equally)* $\dfrac{\zeta'(\sigma)}{\zeta(\sigma)} \ge -\dfrac{1}{\sigma - 1}$.

Proof. We develop an initial formula for $\zeta'(\sigma)$, which we will later modify to suit our needs. In this first pass, we note that, for some $\epsilon > 0$, we have $\sigma = 1 + 2\epsilon$. We start with

$$\zeta'(\sigma) = \frac{d}{ds}\left[\sum_{n=1}^{\infty}\left(\frac{1}{n^\sigma}\right)\right].$$

We know the sum is convergent. Because the sum of the derivatives is also convergent (see equation (16.1) below), we can assume that the derivative of the sum is equal to the sum of the derivatives.

$$= \sum_{n=1}^{\infty}\frac{d}{ds}(n^{-\sigma}) = \sum_{n=1}^{\infty} -n^{-\sigma}\log n = -\sum_{n=1}^{\infty} n^{-\sigma}\log n$$

$$-\zeta'(\sigma) = \sum_{n=1}^{\infty} n^{-\sigma}\log n.$$

To see that the final sum above is convergent we apply lemma 8.1 to our given ϵ. We can then let N be large enough so that, for $n > N$, we have $\log n < n^\epsilon$. Therefore

$$\sum_{n=1}^{\infty} n^{-\sigma}\log n = \sum_{n=1}^{N}\frac{\log n}{n^{1+2\epsilon}} + \sum_{n=N+1}^{\infty}\left[\frac{1}{n^{1+\epsilon}}\cdot\frac{\log n}{n^\epsilon}\right] \le \sum_{n=1}^{N}\frac{\log n}{n^{1+2\epsilon}} + \sum_{n=N+1}^{\infty}\frac{1}{n^{1+\epsilon}} < \infty. \qquad (16.1)$$

Our next step is to use telescoping to convert $-\zeta'(\sigma)$ into a form more useful for our purposes.

$$-\zeta'(\sigma) = \sum_{n=1}^{\infty} n^{-\sigma}\log n = \sum_{n=1}^{\infty}\left[\left(\sum_{m\ge n+1} m^{-\sigma}\right)(\log(n+1) - \log n)\right]$$

$$= \sum_{n=1}^{\infty}\left[\left(\sum_{m\ge n+1} m^{-\sigma}\right)\left(\log\left(1 + \frac{1}{n}\right)\right)\right]. \qquad (16.2)$$

Next, we develop inequalities for the two items in parenthesis in equation (16.2). For the first item, we have a sum of the form $\zeta(\sigma) - \zeta_N(\sigma)$, where ζ_N represents the first N terms of the sum. Because the terms of the sum are non-negative and monotone decreasing, we have (using the integral test for convergence)

$$\int_N^\infty x^{-\sigma}\,dx \le \sum_{n=N}^{\infty} n^{-\sigma} \le \frac{1}{N} + \int_N^\infty x^{-\sigma}\,dx$$

$$\sum_{n=N+1}^{\infty} n^{-\sigma} \le \int_N^\infty x^{-\sigma}\,dx = \frac{N^{1-\sigma}}{\sigma - 1}.$$

For the second item, we can use the Taylor Series for $\log(1+x)$ and obtain $\log(1+x) < 1/x$. Combining we have

$$-\zeta'(\sigma) = \sum_{n=1}^{\infty}\left[\left(\sum_{m\ge n+1} m^{-\sigma}\right)\left(\log\left(1 + \frac{1}{n}\right)\right)\right] \le \sum_{n=1}^{\infty}\left[\left(\frac{n^{1-\sigma}}{\sigma - 1}\right)\left(\frac{1}{n}\right)\right] = \frac{\zeta(\sigma)}{\sigma - 1},$$

so that

$$-\frac{\zeta'(\sigma)}{\zeta(\sigma)} \le \frac{1}{\sigma - 1}.$$

$\square$

Lemma 16.3. *Let $s = \sigma + it$ with $\sigma \ge 1$. We have*

$$\frac{d}{ds} \log \left(\Gamma \left(\frac{s}{2} + 1 \right) \right) = \frac{1}{2} \left[\log \frac{s}{2} + \mathcal{O} \left(\frac{1}{|s|} \right) \right].$$

Proof. We stop to note that Ψ (not to be confused with ψ, the second Chebyshev function) is the digamma function, the logarithmic derivative of the gamma function. We therefore have, applying the chain rule

$$\frac{d}{ds} \log \left(\Gamma \left(\frac{s}{2} + 1 \right) \right) = \frac{1}{2} \Psi \left(\frac{s}{2} + 1 \right).$$

Now using a series expansion of the digamma function, we have

$$\Psi(z + 1) = \log z + \frac{1}{2z} - \sum_{n=1}^{\infty} \frac{B_{2n}}{2nz^{2n}} \quad \text{(where } B_{2n} \text{ are Bernoulli numbers)}$$

$$= \log z + \mathcal{O} \left(\frac{1}{|z|} \right),$$

so that

$$\frac{d}{ds} \log \left(\Gamma \left(\frac{s}{2} + 1 \right) \right) = \frac{1}{2} \Psi \left(\frac{s}{2} + 1 \right) = \frac{1}{2} \left[\log \frac{s}{2} + \frac{1}{s} - \sum_{n=1}^{\infty} \frac{B_{2n}}{2n(s/2)^{2n}} \right] = \frac{1}{2} \left[\log \frac{s}{2} + \mathcal{O} \left(\frac{1}{|s|} \right) \right]. \quad \square$$

Lemma 16.4. *Let $\{s = \sigma + it \in \mathbb{C} : 1 \le \sigma \le 2\}$. Then there is a $K > 1$ such that for $t > K$ we have*

$$Re \left[\frac{\zeta'(\sigma + it)}{\zeta(\sigma + it)} \right] > Re \left[\frac{1}{\sigma + it - \rho} \right] - \log t + \frac{1}{2} \log \pi.$$

Proof. We start with equation (12.1), where ρ has its usual meaning as the roots of $\xi(s)$ (see page xi):

$$\frac{\zeta'(s)}{\zeta(s)} = \sum_{\rho} \frac{1}{s - \rho} - \left[\frac{d}{ds} \log \left(\Gamma \left(\frac{s}{2} + 1 \right) \right) \right] + \frac{1}{2} \log \pi - \frac{1}{s - 1}.$$

Using lemma 16.3, we substitute for the formula in brackets

$$= \sum_{\rho} \frac{1}{s - \rho} - \frac{1}{2} \left[\log \frac{s}{2} + \mathcal{O} \left(\frac{1}{|s|} \right) \right] + \frac{1}{2} \log \pi - \frac{1}{s - 1}. \tag{16.3}$$

Before we return to equation (16.3) we do two simple calculations:

$$0 \le Re \left(\frac{1}{s - 1} \right) = \frac{\sigma - 1}{(\sigma - 1)^2 + t^2} < \frac{1}{t^2} \quad \text{(because } t > 1 \text{ and } 1 \le \sigma \le 2\text{).} \tag{16.4}$$

Noting that $Re(\log s) = \log |s| = \log \sqrt{\sigma^2 + t^2} = 1/2 \log(\sigma^2 + t^2)$, we also have

$$0 < Re \left(\frac{1}{2} \log \frac{s}{2} \right) = \frac{1}{4} \log \left(\sigma^2 + t^2 \right) - \frac{\log 2}{2} \le \frac{1}{4} \log \left(4t^2 \right) - \frac{\log 2}{2} = \frac{1}{2} \log t. \tag{16.5}$$

Returning to equation (16.3), we use equations (16.4) and (16.5), and we have for some $C > 1$ (used for the $\mathcal{O}(1/|s|)$ term)

$$Re\left[\frac{\zeta'(\sigma + it)}{\zeta(\sigma + it)}\right] \geq \sum_\rho Re\left(\frac{1}{s - \rho}\right) + \frac{1}{2}\log\pi - \frac{1}{2}\log t - \frac{C}{t} - \frac{1}{t^2}$$

$$= \sum_\rho Re\left(\frac{1}{s - \rho}\right) + \frac{1}{2}\log\pi - \log t + \left[\frac{1}{2}\log t - \frac{C}{t} - \frac{1}{t^2}\right].$$

Now assume $t > K > \max(C, e^4)$, so that

$$Re\left[\frac{\zeta'(\sigma + it)}{\zeta(\sigma + it)}\right] \geq \sum_\rho Re\left(\frac{1}{s - \rho}\right) + \frac{1}{2}\log\pi - \log t + \left[\frac{1}{2}\log e^4 - \frac{C}{C} - \frac{1}{C^2}\right]$$

$$> \sum_\rho Re\left(\frac{1}{s - \rho}\right) + \frac{1}{2}\log\pi - \log t$$

$$= \sum_\rho Re\left(\frac{1}{\sigma + it - \rho}\right) + \frac{1}{2}\log\pi - \log t.$$

Because $1 \leq \sigma \leq 2$ and $0 < Re(\rho) < 1$, we have $\sigma - Re(\rho) > 0$. That means the real part of each term in the sum is positive. Therefore, the inequality still holds if we omit one or more terms of the sum. In particular, we can include only one term and have (for $t > K$)

$$Re\left[\frac{\zeta'(\sigma + it)}{\zeta(\sigma + it)}\right] > Re\left[\frac{1}{\sigma + it - \rho}\right] - \log t + \frac{1}{2}\log\pi. \qquad \square$$

16.1.2 Proof of Zero-Free Region

Theorem 16.1. *Let $\rho = \beta + i\gamma \in \mathbb{C}$. Then there exist constants $c > 0$, $K > 1$, such that*

$$\beta \leq 1 - \frac{c}{\log\gamma}$$

in all cases where $\zeta(\rho) = 0$ and $\gamma > K$.

Proof. We assume throughout that $1 < \sigma \leq 2$, and begin with the results from lemma 16.1

$$0 \geq Re\left[3\frac{\zeta'(\sigma)}{\zeta(\sigma)}\right] + Re\left[4\frac{\zeta'(\sigma + it)}{\zeta(\sigma + it)}\right] + Re\left[\frac{\zeta'(\sigma + 2it)}{\zeta(\sigma + 2it)}\right].$$

Now consider the three terms individually. For the first term, we have from lemma 16.2

$$Re\left[3\frac{\zeta'(\sigma)}{\zeta(\sigma)}\right] \geq -\frac{3}{\sigma - 1}.$$

For the second and third terms, we have from lemma 16.4

$$Re\left[4\frac{\zeta'(\sigma + it)}{\zeta(\sigma + it)}\right] > Re\left[\frac{4}{\sigma + it - \rho}\right] - 4\log t + 2\log\pi$$

$$Re\left[\frac{\zeta'(\sigma + 2it)}{\zeta(\sigma + 2it)}\right] > Re\left[\frac{1}{\sigma + 2it - \rho}\right] - \log(2t) + \frac{1}{2}\log\pi.$$

Putting the three terms back together gives

$$0 \geq -\frac{3}{\sigma - 1} + Re\left[\frac{4}{\sigma + it - \rho}\right] - 4\log t + Re\left[\frac{1}{\sigma + 2it - \rho}\right] - \log(2t) + \frac{5}{2}\log \pi.$$

Because $\sigma > Re(\rho)$, the second bracketed term is positive and can be disregarded. We also use the fact that $-\log(2t) = -\log 2 - \log t$.

$$0 \geq -\frac{3}{\sigma - 1} + Re\left[\frac{4}{\sigma + it - \rho}\right] - 5\log t + \left[-\log 2 + \frac{5}{2}\log \pi\right].$$

The second bracketed term is positive and can be disregarded. We also use $C > 0$ to represent our multiple of $\log t$, giving

$$0 \geq -\frac{3}{\sigma - 1} + Re\left[\frac{4}{\sigma + it - \rho}\right] - C\log t.$$

We therefore have, for all roots ρ with $t > K$, where $1 < \sigma \leq 2$

$$\frac{3}{\sigma - 1} - Re\left[\frac{4}{\sigma + it - \rho}\right] \geq -C\log t.$$

Note that C is independent of σ, t and ρ. Our only requirement is that our large K be large enough so that $\log K > 1/C$. Now let a root $\rho = \beta + i\gamma$ be given, with $\gamma > K$. We can therefore set $t = \gamma$ and have[39]

$$\frac{3}{\sigma - 1} - \frac{4}{\sigma - \beta} \geq -C\log t$$

$$-\frac{4}{\sigma - \beta} \geq -\left[C\log t + \frac{3}{\sigma - 1}\right]$$

$$\frac{4}{\sigma - \beta} \leq C\log t + \frac{3}{\sigma - 1} = \frac{3 + (\sigma - 1)C\log t}{(\sigma - 1)}$$

$$\sigma - \beta \geq \frac{4(\sigma - 1)}{3 + (\sigma - 1)C\log t}$$

$$(\sigma - 1) + 1 - \beta \geq \frac{4(\sigma - 1)}{3 + (\sigma - 1)C\log t}$$

$$1 - \beta \geq \frac{4(\sigma - 1)}{3 + (\sigma - 1)C\log t} - (\sigma - 1) = \frac{4(\sigma - 1)}{3 + (\sigma - 1)C\log t} \cdot \frac{(\sigma - 1)^{-1}}{(\sigma - 1)^{-1}} - (\sigma - 1)$$

$$\geq \frac{4}{3/(\sigma - 1) + C\log t} - (\sigma - 1)$$

$$\geq \frac{4}{3/(\sigma - 1) + C\log t} - \frac{(\sigma - 1) \cdot [3/(\sigma - 1) + C\log t]}{3/(\sigma - 1) + C\log t}$$

$$\geq \frac{1 - (\sigma - 1)C\log t}{3/(\sigma - 1) + C\log t}.$$

Now[40] set $(\sigma - 1) = 1/[2C \cdot \log t]$. (We have kept σ in the range $1 < \sigma \leq 2$ because $\log t > \log K$ and $C \cdot \log K > 1$). We also set $c = (14 \cdot C)^{-1}$ and obtain our result as follows

$$1 - \beta \geq \frac{1 - (1/[2C \cdot \log t])C\log t}{3 \cdot [2C \cdot \log t] + C\log t} = \frac{\frac{1}{2}}{7C\log t} = \frac{c}{\log t}$$

$$\beta \leq 1 - \frac{c}{\log t}. \qquad \square$$

16.2 An Error Function and Error Term for $\psi(x) \sim x$

In this section, we use the zero-free region established in Theorem 16.1 to obtain an error function (and therefore an error term) for $\psi(x) \sim x$. Almost all of the work in obtaining our error function is done in the first two lemmas immediately below.[41]

Lemma 16.5. *Let* $F(\gamma) = \dfrac{x^{-c/\log \gamma}}{\gamma^{1-\delta}}$. *Then, the logarithmic derivative of* $F(\gamma)$ *is*

$$\frac{F'(\gamma)}{F(\gamma)} = \frac{c \log x}{\gamma (\log \gamma)^2} - \frac{(1-\delta)}{\gamma}.$$

Proof. Set $g(\gamma) = x^{-c/\log \gamma}$ and $h(\gamma) = \gamma^{1-\delta}$, so that $F(\gamma) = g(\gamma)/h(\gamma)$ and therefore

$$F'(\gamma) = \frac{g'(\gamma)h(\gamma) - g(\gamma)h'(\gamma)}{h^2(\gamma)}.$$

We compute $g'(\gamma)$ by the chain rule. Set $q(\gamma) = -c \log x / \log \gamma$. Then

$$g(\gamma) = x^{-c/\log \gamma} = e^{q(\gamma)} \quad \text{so that} \quad g'(\gamma) = \frac{d}{d\gamma}e^{q(\gamma)} = e^{q(\gamma)}q'(\gamma)$$

$$q'(\gamma) = -c \log x \left(\frac{d}{d\gamma}\frac{1}{\log \gamma} \right) = -c \log x \left(\frac{-1}{\gamma(\log \gamma)^2} \right) = \frac{c \log x}{\gamma(\log \gamma)^2}$$

$$g'(\gamma) = e^{-c \log x / \log \gamma}\frac{c \log x}{\gamma(\log \gamma)^2}.$$

It is easily seen that $h'(\gamma) = (1-\delta)\gamma^{-\delta}$, giving

$$F'(\gamma) = \frac{g'(\gamma)h(\gamma) - g(\gamma)h'(\gamma)}{h^2(\gamma)}$$

$$= \frac{e^{-c \log x / \log \gamma}\dfrac{c \log x}{\gamma(\log \gamma)^2}\gamma^{1-\delta} - e^{-c \log x / \log \gamma}(1-\delta)\gamma^{-\delta}}{\gamma^{1-\delta}\gamma^{1-\delta}}$$

$$= \frac{e^{-c \log x / \log \gamma}\dfrac{c \log x}{\gamma(\log \gamma)^2}\gamma^{1-\delta}}{\gamma^{1-\delta}\gamma^{1-\delta}} - \frac{e^{-c \log x / \log \gamma}(1-\delta)\gamma^{-\delta}}{\gamma^{1-\delta}\gamma^{1-\delta}}$$

$$= \frac{e^{-c \log x / \log \gamma}c \log x}{\gamma(\log \gamma)^2\gamma^{1-\delta}} - \frac{e^{-c \log x / \log \gamma}(1-\delta)\gamma^{-\delta}}{\gamma^{1-\delta}\gamma^{1-\delta}}.$$

Putting it all together, we have

$$\frac{F'(\gamma)}{F(\gamma)} = \frac{F'(\gamma)\gamma^{1-\delta}}{e^{-c \log x / \log \gamma}}$$

$$= \frac{\gamma^{1-\delta}}{e^{-c \log x / \log \gamma}}\left[\frac{e^{-c \log x / \log \gamma}c \log x}{\gamma(\log \gamma)^2\gamma^{1-\delta}} - \frac{e^{-c \log x / \log \gamma}(1-\delta)\gamma^{-\delta}}{\gamma^{1-\delta}\gamma^{1-\delta}} \right]$$

$$= \frac{c \log x}{\gamma(\log \gamma)^2} - \frac{(1-\delta)}{\gamma}. \qquad \square$$

Lemma 16.6. *For x sufficiently large and for some c > 0*

$$\left| \frac{\int_0^x \psi(t)\, dt - \frac{x^2}{2}}{\frac{x^2}{2}} \right| \le e^{-[c \log x]^{1/2}}.$$

Proof. From lemma 14.5, our goal is to obtain a formula for the "error function" as discussed in Section 12.1. That is, the rate at which the below limit approaches 0:

$$\lim_{x \to \infty} \sum_\rho \frac{x^{\rho-1}}{\rho(\rho+1)} = 0.$$

We will use our zero-free region established in Theorem 16.1 to obtain that formula.

Let $\{\rho = \beta + i\gamma \in \mathbb{C} : 0 < \beta < 1\}$ and let $0 < \delta < 1$. We have (choosing K as described further below):

$$\left| \sum_\rho \frac{x^{\rho-1}}{\rho(\rho+1)} \right| \le \sum_\rho \frac{x^{\beta-1}}{\gamma^2} = \sum_{|\gamma|<K} \frac{x^{\beta-1}}{\gamma^2} + 2 \sum_{\gamma \ge K} \frac{x^{\beta-1}}{\gamma^{1-\delta}} \cdot \frac{1}{\gamma^{1+\delta}}. \tag{16.6}$$

Consider the first sum on the RHS. For our finite number of terms of the form $\rho = \beta + i\gamma$, let ϵ equal the minimum of the $(1 - \beta)$ values. Note that $\epsilon > 0$. Then, for some $C_1 > 0$ (and for all $x > 1$)

$$\sum_{|\gamma|<K} \frac{x^{\beta-1}}{\gamma^2} \le \sum_{|\gamma|<K} \frac{1}{x^\epsilon \gamma^2} = x^{-\epsilon} \sum_{|\gamma|<K} \frac{1}{\gamma^2} = C_1 x^{-\epsilon}.$$

Now consider the second sum on the RHS. First notice that we multiplied the result by 2 so that we could remove the absolute value around γ and assume $\gamma \ge K$. We choose K from lemma 9.7 and therefore know that the sum (if it were just over the second multiplicand) converges – we will assign that sum the fixed value C_2. Assuming that the first multiplicand takes on a maximum value, then the sum still converges, and is less than C_2 times the maximum value of the first multiplicand.

We now study that first multiplicand. We can use Theorem 16.1 to obtain

$$\frac{x^{\beta-1}}{\gamma^{1-\delta}} \le \frac{x^{-c/\log \gamma}}{\gamma^{1-\delta}}.$$

It is a bit of a laborious calculation (see lemma 16.5), but the logarithmic derivative of the RHS, taken with respect to γ, is

$$\frac{c \log x}{\gamma (\log \gamma)^2} - \frac{(1-\delta)}{\gamma} = \frac{c \log x}{\gamma (\log \gamma)^2} - \frac{(1-\delta)(\log \gamma)^2}{\gamma (\log \gamma)^2}.$$

Remembering that $\gamma \ge K$, we see that the sign of the result is based on the relative values of $c \log x$ and $(1-\delta)(\log \gamma)^2$. For our fixed K, let x be large enough so that $c \log x > (1 - \delta)(\log K)^2$. Thus, we reach a maximum for $\gamma \ge K$ at the point where $c \log x = (1 - \delta)(\log \gamma)^2$. (The peak between a positive or negative slope). At that point

$$\frac{c \log x}{\log \gamma} = (1-\delta) \log \gamma = \sqrt{(1-\delta)^2 (\log \gamma)^2} = \sqrt{(1-\delta) \log \gamma \frac{c \log x}{\log \gamma}} = \sqrt{(1-\delta) c \log x}.$$

We therefore have (setting $\delta = \frac{3}{4}$)

$$\frac{x^{\beta-1}}{\gamma^{1-\delta}} \leq \frac{x^{-c/\log\gamma}}{\gamma^{1-\delta}} = \frac{e^{\frac{-c\log x}{\log\gamma}}}{e^{(1-\delta)\log\gamma}} = \frac{e^{-\sqrt{(1-\delta)c\log x}}}{e^{\sqrt{(1-\delta)c\log x}}} = \frac{e^{-\sqrt{1/4}\sqrt{c\log x}}}{e^{\sqrt{1/4}\sqrt{c\log x}}} = \frac{e^{-\frac{1}{2}\sqrt{c\log x}}}{e^{\frac{1}{2}\sqrt{c\log x}}} = e^{-[c\log x]^{1/2}}.$$

Returning to equation (16.6), we have

$$\left| \sum_\rho \frac{x^{\rho-1}}{\rho(\rho+1)} \right| \leq \sum_{|\gamma|<K} \frac{x^{\beta-1}}{\gamma^2} + 2\sum_{\gamma\geq K} \frac{x^{\beta-1}}{\gamma^{1-\delta}} \cdot \frac{1}{\gamma^{1+\delta}} \leq C_1 x^{-\epsilon} + C_2 e^{-[c\log x]^{1/2}}$$

for sufficiently large x. Because C_1 and C_2 are fixed, and $x^{-\epsilon}$ decays much faster than $e^{-[c\log x]^{1/2}}$, we can decrease c as necessary to have

$$\left| \sum_\rho \frac{x^{\rho-1}}{\rho(\rho+1)} \right| \leq e^{-[c_1 \log x]^{1/2}}.$$

For the same reasons, the other terms in equation (14.2) can be absorbed by an adustment to c_1, so that the relative error in the asymptotic $\int_0^\infty \psi(t)\, dt \sim x^2/2$ is $\leq e^{-[c_2 \log x]^{1/2}}$ for sufficiently large x. □

Theorem 16.2. *For x sufficiently large and for some $c > 0$*

$$\left| \frac{\psi(x) - x}{x} \right| \leq e^{-[c\log x]^{1/2}} \qquad \text{and therefore} \qquad |\psi(x) - x| \leq \frac{x}{e^{[c\log x]^{1/2}}}.$$

Proof. Fix $K > 0$ such that, from lemma 16.6, we have for $x > K$ and for some $c > 0$

$$\left| \frac{\int_0^x \psi(t)\, dt - \frac{x^2}{2}}{\frac{x^2}{2}} \right| \leq e^{-[c\log x]^{1/2}}.$$

Now apply lemma 14.7, with $h(x) = e^{-[c\log x]^{1/2}}$ and $\epsilon = \sqrt{h(x)}$, so that for $x > K$ we have

$$\frac{\psi(x) - x}{x} \leq \frac{\epsilon}{2} + h(x) + \frac{h(x)\epsilon}{2} + \frac{h(x)}{\epsilon} \tag{16.7}$$

$$\frac{\psi(x) - x}{x} \geq -\frac{\epsilon}{2} - h(x) - \frac{h(x)}{\epsilon} \tag{16.8}$$

We can (and will) assume that $h(x) < 1$, so that $0 < h(x) < \epsilon = \sqrt{h(x)} < 1$.
For equation (16.7), we have

$$\frac{\psi(x) - x}{x} \leq \frac{\sqrt{h(x)}}{2} + h(x) + \frac{h(x)\sqrt{h(x)}}{2} + \frac{h(x)}{\sqrt{h(x)}}$$

$$\boxed{\frac{\psi(x) - x}{x}} \leq \frac{\sqrt{h(x)}}{2} + \sqrt{h(x)} + \frac{\sqrt{h(x)}}{2} + \sqrt{h(x)} \boxed{\leq 3\sqrt{h(x)}}.$$

For equation (16.8), we have

$$\frac{\psi(x) - x}{x} \geq -\frac{\sqrt{h(x)}}{2} - h(x) - \frac{h(x)}{\sqrt{h(x)}}$$

$$\boxed{\frac{\psi(x) - x}{x}} \geq -\frac{\sqrt{h(x)}}{2} - \sqrt{h(x)} - \sqrt{h(x)} \boxed{\geq -3\sqrt{h(x)}}$$

Now note that $\sqrt{h(x)} = \sqrt{e^{-[c \log x]^{1/2}}} = \left(e^{-[c \log x]^{1/2}}\right)^{1/2} = e^{-\frac{1}{2}[c \log x]^{1/2}}$ Putting the above together, we have our *error function* and *error term*, as described in section 12.1:

$$\left|\frac{\psi(x) - x}{x}\right| \leq 3\sqrt{h(x)} \leq e^{-[c_2 \log x]^{1/2}} \quad \text{and therefore} \quad |\psi(x) - x| \leq \frac{x}{e^{[c_2 \log x]^{1/2}}},$$

for some $c_2 > 0$ and x sufficiently large. $\qquad \square$

16.3 Error Terms and the Riemann Hypothesis

We showed in section 16.2 that there is a close relationship between the zero-free region of $\zeta(s)$ in the critical strip and the error term in the asymptotic functions associated with the *Prime Number Theorem*. We can take that thought further by assuming the Riemann Hypothesis is true and thereby expanding the zero-free region to its maximum. That is, we assume $\zeta(\beta + i\gamma) \neq 0$ for $\beta > 1/2$.

We will consider the question in both directions. First, we will assume the Riemann Hypothesis is true and obtain a much tighter error term for $|\psi(x) - x|$. Then we will assume a certain error term in $|\psi(x) - x|$ and show, if true, the Riemann Hypothesis must also be true.[42]

16.3.1 Two Useful Lemmas

Lemma 16.7. *For $\theta(x)$ and $\psi(x)$, we have*

$$\psi(x) \leq \theta(x) + \frac{\theta(x^{1/2}) \log x}{\log 2}.$$

Proof. We state various properties of $\theta(x)$ and $\psi(x)$:

$$\theta(x) = \sum_{p \leq x} \log p \quad \Longrightarrow \quad \theta(x) \leq x \log x \quad \Longrightarrow \quad \theta(x^{1/m}) \leq x^{1/m} \log x \quad (\text{for } x, m \geq 2),$$

$$\psi(x) = \sum_{p^n \leq x} \log p = \sum_{k=1}^{\infty} \theta(x^{1/k}) = \sum_{k=1}^{\infty} \left(\sum_{p \leq x^{1/k}} \log p\right) \quad (\text{because } p^k \leq x \Rightarrow p \leq x^{1/k}).$$

Note that $x^{1/k} < 2$ means $\theta(x^{1/k}) = 0$. Equally, $\log(x^{1/k}) < \log 2$ means $\theta(x^{1/k}) = 0$. Therefore, if we assume that $k > \log x / \log 2$, then $x^{1/k} < x^{\log 2 / \log x} < e^{\log 2} < 2$. We can therefore define $M(x)$ as follows and have

$$M(x) = \left[\frac{\log x}{\log 2}\right] \quad \Longrightarrow \quad \psi(x) = \sum_{k=1}^{M(x)} \theta(x^{1/k}).$$

Continuing with our defined $M(x)$

$$\psi(x) = \theta(x) + \sum_{k=2}^{M(x)} \theta(x^{1/k}) \le \theta(x) + \left[M(x) \cdot \left(\theta(x^{1/2})\right)\right] \le \theta(x) + \frac{\theta(x^{1/2})\log x}{\log 2}. \qquad \square$$

Lemma 16.8. *For $x \ge 2$ we have*

$$\theta(x) = \pi(x)\log x - \int_2^x \frac{\pi(t)}{t}\,dt,$$

$$\pi(x) = \frac{\theta(x)}{\log x} + \int_2^x \frac{\theta(t)}{t \cdot (\log t)^2}\,dt.$$

Proof. Define the characteristic function $\chi(n)$ to be 1 if n is prime and 0 otherwise, so that

$$\pi(x) = \sum_{1 \le n \le x} \chi(n) \quad \text{and} \quad \theta(x) = \sum_{1 \le n \le x} \chi(n)\log n.$$

For $\theta(x)$, apply lemma 5.3, setting $f(n) = \chi(n)$, $g(x) = \log x$ and $y = 1$, to obtain

$$\sum_{y \le n \le x} \chi(n)\log n = \sum_{1 \le n \le x} \chi(n)\cdot \log x - 0 - \int_1^x \left(\sum_{1 \le n \le t}\chi(t)\right)\cdot \frac{1}{t}\,dt$$

$$\theta(x) = \pi(x)\log x - \int_2^x \frac{\pi(t)}{t}\,dt.$$

For $\pi(x)$, apply lemma 5.3, setting $f(n) = \chi(n)\log n$, $g(x) = 1/\log x$ and $y = 1.5$, to obtain

$$\sum_{y \le n \le x} \frac{\chi(n)\log n}{\log n} = \left(\sum_{1 \le n \le x}\chi(n)\log n\right)\cdot \frac{1}{\log x} - 0 - \int_{1.5}^x \left(\sum_{1 \le n \le t}\chi(t)\log t\right)\cdot \frac{-1}{t\cdot(\log t)^2}\,dt$$

$$\pi(x) = \frac{\theta(x)}{\log x} + \int_{1.5}^2 \frac{\theta(t)}{t\cdot(\log t)^2}\,dt + \int_2^x \frac{\theta(t)}{t\cdot(\log t)^2}\,dt = \frac{\theta(x)}{\log x} + \int_2^x \frac{\theta(t)}{t\cdot(\log t)^2}\,dt. \qquad \square$$

16.3.2 Assume the Riemann Hypothesis – Obtain Error Term

Theorem 16.3. *Assume the Riemann Hypothesis is true, so that $\zeta(\beta + i\gamma) \ne 0$ for $\beta > 1/2$. Then, For x sufficiently large and for some $c > 0$*

$$
\begin{array}{lll}
(i) & |\psi(x) - x| & < cx^{1/2}(\log x)^2 \\
(ii) & |\theta(x) - x| & < cx^{1/2}(\log x)^2 \\
(iii) & |\pi(x) - Li(x)| & < cx^{1/2}(\log x).
\end{array}
$$

NOTE: In all that follows, $C > 0$ represents the appropriate constant value in context, and does not represent any particular fixed constant value.

Proof – (i). Because $\psi(x)$ is monotone increasing, we have

$$\psi(x) \le \int_x^{x+1} \psi(t)\,dt = \int_0^{x+1}\psi(t)\,dt - \int_0^x \psi(t)\,dt$$

$$\psi(x) \ge \int_{x-1}^x \psi(t)\,dt = \int_0^x \psi(t)\,dt - \int_0^{x-1}\psi(t)\,dt.$$

Combined with the results from lemma 14.4 (repeated here for convenience)

$$\int_0^x \psi(t)\,dt = \frac{x^2}{2} - \sum_\rho \frac{x^{\rho+1}}{\rho(\rho+1)} - \sum_{n=1}^\infty \frac{x^{-2n+1}}{2n(2n-1)} - x\frac{\zeta'(0)}{\zeta(0)} + \frac{\zeta'(-1)}{\zeta(-1)} \quad (x > 1),$$

we have

$$\psi(x) \le \left| \frac{(x+1)^2}{2} - \frac{x^2}{2} \right| + \left| \sum_\rho \frac{(x+1)^{\rho+1} - x^{\rho+1}}{\rho(\rho+1)} \right| + C$$

$$\psi(x) \le x + \left| \sum_\rho \frac{(x+1)^{\rho+1} - x^{\rho+1}}{\rho(\rho+1)} \right| + C$$

and similarly

$$\psi(x) \ge \left| \frac{x^2}{2} - \frac{(x-1)^2}{2} \right| - \left| \sum_\rho \frac{x^{\rho+1} - (x-1)^{\rho+1}}{\rho(\rho+1)} \right| - C$$

$$\psi(x) \ge x - \left| \sum_\rho \frac{x^{\rho+1} - (x-1)^{\rho+1}}{\rho(\rho+1)} \right| - C.$$

By hypothesis, all root are of the form $\rho = 1/2 + i\gamma$. We divide those root ρ as follows.
For roots ρ with $|\gamma| \le x$, we have (using ML inequality)

$$\left| \frac{(x+1)^{\rho+1} - x^{\rho+1}}{\rho(\rho+1)} \right| = \frac{1}{|\rho|} \left| \int_x^{x+1} t^\rho\,dt \right| \le \frac{|x+1|^{1/2}}{|\rho|} \le \frac{(2x)^{1/2}}{|\gamma|} = x^{1/2} \cdot \frac{\sqrt{2}}{|\gamma|}$$

$$\left| \frac{x^{\rho+1} - (x-1)^{\rho+1}}{\rho(\rho+1)} \right| = \frac{1}{|\rho|} \left| \int_{x-1}^x t^\rho\,dt \right| \le \frac{|x|^{1/2}}{|\rho|} \le \frac{x^{1/2}}{|\gamma|}$$

and for roots ρ with $|\gamma| > x$, we have

$$\left| \frac{(x+1)^{\rho+1} - x^{\rho+1}}{\rho(\rho+1)} \right| \le \frac{2 \cdot |x+1|^{3/2}}{\rho(\rho+1)} \le \frac{2 \cdot (2x)^{3/2}}{\gamma^2} = x^{3/2} \cdot \frac{2^{5/2}}{\gamma^2}$$

$$\left| \frac{x^{\rho+1} - (x-1)^{\rho+1}}{\rho(\rho+1)} \right| \le \frac{2 \cdot |x|^{3/2}}{\rho(\rho+1)} \le \frac{2 \cdot x^{3/2}}{\gamma^2}.$$

Applying Theorem 9.1 (see also lemma 9.1), there is a $K > 0$ such that, for $x > K$, there are at most $2\log t$ roots $\rho = 1/2 + i\gamma$ in the interval $t \le \gamma \le t + 1$. We will assume that $x > K$, and consider the three possible cases for the value of γ: (1) $|\gamma| \le K$, (2) $\gamma > K$ and $\gamma \le x$, and (3) $\gamma > K$ and $\gamma > x$. In the last two cases, we multiply the result by 2 to account for the positive and negative roots with γ.
Now establish a maximum value for $\psi(x)$.

$$\psi(x) \le x + \left(x^{1/2} \sum_{|\gamma| \le K} \frac{\sqrt{2}}{|\gamma|} \right) + \left(2x^{1/2} \sum_{K < \gamma \le x} \frac{\sqrt{2}}{\gamma} \right) + \left(2x^{3/2} \sum_{K < x < \gamma} \frac{2^{5/2}}{\gamma^2} \right)$$

$$\le x + \left(C \cdot x^{1/2} \right) + \left(C \cdot x^{1/2} \cdot \int_K^x \frac{\log t}{t}\,dt \right) + \left(C \cdot x^{3/2} \cdot \int_x^\infty \frac{\log t}{t^2}\,dt \right)$$

$$\le x + \left(C \cdot x^{1/2} \right) + \left(C \cdot x^{1/2} \cdot \frac{(\log t)^2}{2} \Big|_H^x \right) + \left(C \cdot x^{3/2} \cdot \left[\frac{\log x}{x} + \frac{1}{x} \right] \right)$$

$$\le x + \left(C \cdot x^{1/2} \right) + \left(C \cdot x^{1/2} \cdot (\log x)^2 \right) + \left(C \cdot x^{1/2} \cdot [\log x + 1] \right)$$

$$\psi(x) \le x + C \cdot x^{1/2} \cdot (\log x)^2.$$

Finally, establish a minimum value for $\psi(x)$.

$$\psi(x) \geq x - \left(x^{1/2} \sum_{|\gamma| \leq K} \frac{1}{|\gamma|}\right) - \left(2x^{1/2} \sum_{K < \gamma \leq x} \frac{1}{\gamma}\right) - \left(4x^{3/2} \sum_{K < x < \gamma} \frac{1}{\gamma^2}\right)$$

$$\geq x - \left(C \cdot x^{1/2}\right) - \left(C \cdot x^{1/2} \cdot \int_K^x \frac{\log t}{t}\, dt\right) - \left(C \cdot x^{3/2} \cdot \int_x^\infty \frac{\log t}{t^2}\, dt\right)$$

$$\geq x - \left(C \cdot x^{1/2}\right) - \left(C \cdot x^{1/2} \cdot \frac{(\log t)^2}{2}\Big|_H^x\right) - \left(C \cdot x^{3/2} \cdot \left[\frac{\log x}{x} + \frac{1}{x}\right]\right)$$

$$\geq x - \left(C \cdot x^{1/2}\right) - \left(C \cdot x^{1/2} \cdot (\log x)^2\right) - \left(C \cdot x^{1/2} \cdot [\log x + 1]\right)$$

$$\psi(x) \geq x - C \cdot x^{1/2} \cdot (\log x)^2.$$

Combining our results, we have $|\psi(x) - x| \leq C x^{1/2} (\log x)^2$. $\qquad\square$

Proof – (ii). We assume the results from (i), immediately above. Using lemma 16.7, we have

$$\psi(x) - \frac{\theta(x^{1/2}) \log x}{\log 2} \leq \theta(x) \leq \psi(x).$$

Because $\theta(x) \sim x$, we also have $\theta(x^{1/2}) \log x \sim x^{1/2} \log x$. But $x^{1/2} \log x < x^{1/2} (\log x)^2$. That means, for sufficiently large x

$$\psi(x) - C x^{1/2} (\log x)^2 \leq \theta(x) \leq \psi(x). \qquad\square$$

Proof – (iii). We assume the results from (i) and (ii), immediately above. Using integration by parts, we put $Li(x)$ in the form we need:

$$Li(x) = \int_2^x \frac{dt}{\log t} = \frac{t}{\log t}\Big|_2^x + \int_2^x \frac{dt}{(\log t)^2} = \frac{x}{\log x} - \frac{2}{\log 2} + \int_2^x \frac{dt}{(\log t)^2}.$$

We therefore have (using our formula for $\pi(x)$ in lemma 16.8)

$$\pi(x) - Li(x) = \left[\frac{\theta(x)}{\log x} + \int_2^x \frac{\theta(t)}{t \cdot (\log t)^2}\, dt\right] - \left[\frac{x}{\log x} - \frac{2}{\log 2} + \int_2^x \frac{dt}{(\log t)^2}\right]$$

$$= \frac{\theta(x) - x}{\log x} + \int_2^x \frac{\theta(t) - t}{t \cdot (\log t)^2}\, dt + C.$$

Working now with the numerator in the integrand, we use proof (ii), above, to obtain a $K > 0$ and a constant C such that for $t \geq K$ we have $|\theta(t) - t| < C t^{1/2} (\log t)^2$. Thus, for $x > K$

$$|\pi(x) - Li(x)| \leq \frac{C x^{1/2} (\log x)^2}{\log x} + \int_2^K \frac{\theta(t) - t}{t \cdot (\log t)^2}\, dt + \int_K^x \frac{C t^{1/2} (\log t)^2}{t \cdot (\log t)^2}\, dt + C$$

$$\leq C \frac{x}{\log x} \frac{(\log x)^2}{x^{1/2}} + C + \int_K^x \frac{C}{t^{1/2}}\, dt + C$$

$$\leq C \frac{x}{\log x} \frac{(\log x)^2}{x^{1/2}} + C x^{1/2} + C \leq C x^{1/2} \log x + C x^{1/2} + C$$

$$\leq C x^{1/2} \log x. \qquad\square$$

In Theorem 16.3, just above, we obtained error terms of the form $Cx^{1/2}(\log x)^2$ for $|\psi(x) - x|$ and $|\theta(x) - x|$, and of the form $Cx^{1/2}(\log x)$ for $|\pi(x) - Li(x)|$. Note that the error term can be stated in the form $x^{1/2+\epsilon}$ for any $\epsilon > 0$ for sufficiently large x. (A portion of the ϵ can absorb the constant C and another portion can use lemma 8.1 to absorb the log or $\log^2$ term).

We can also determine the *error function*, as described in section 12.1. In the case of $|\psi(x) - x|$ and $|\theta(x) - x|$, it is simply a matter of dividing the error rate by x, giving $x^{-(1/2)+\epsilon}$. In the case of $|\pi(x) - Li(x)|$, we use $Li(x) \sim x/\log x$ for large enough x and divide the error rate by $x/\log x$

$$\frac{|\pi(x) - Li(x)|}{Li(x)} \le \frac{Cx^{1/2}\log x}{\frac{x}{\log x}} = Cx^{1/2}\log x \cdot \frac{\log x}{x} = Cx^{-(1/2)}(\log x)^2,$$

allowing the same $x^{-(1/2)+\epsilon}$.

16.3.3 Assume Error Term – Prove Riemann Hypothesis

Theorem 16.4. *Assume that for all $\epsilon > 0$ there is a $K > 0$ (which may be dependent on ϵ) such that*

$$|\psi(x) - x| \le x^{(1/2)+\epsilon}$$

for all $x > K$. Then the Riemann Hypothesis is true.

Proof. We start with two equalities, both valid for $Re(s) > 1$:

$$-\frac{\zeta'(s)}{\zeta(s)} = s\int_1^\infty \psi(x)x^{-s-1}\,dx \quad \text{(from lemma 12.1)}$$

$$-\frac{1}{(1-s)} = 1 + s\int_1^\infty x^{-s-1}x\,dx.$$

Combining the two equations, we have (still requiring $Re(s) > 1$)

$$-\frac{\zeta'(s)}{\zeta(s)} - \left[\frac{1}{(s-1)}\right] = s\int_1^\infty \psi(x)x^{-s-1}\,dx - \left[-1 + s\int_1^\infty x^{-s-1}\cdot x\,dx\right]$$

$$-\frac{\zeta'(s)}{\zeta(s)} - \left[\frac{d}{ds}\log(s-1)\right] = 1 + s\int_1^\infty [\psi(x) - x]x^{-s-1}\,dx$$

$$-\frac{d}{ds}\log[(s-1)\zeta(s)] = 1 + s\int_1^\infty [\psi(x) - x]x^{-s-1}\,dx. \tag{16.9}$$

By assumption, $|\psi(x) - x| \le x^{(1/2)+\epsilon}$ for $x > K$. We therefore have (for some constant C)

$$1 + s\int_1^\infty [\psi(x) - x]x^{-s-1}\,dx = 1 + s\int_1^K [\psi(x) - x]x^{-s-1}\,dx + s\int_K^\infty [\psi(x) - x]x^{-s-1}\,dx$$

$$= C + s\int_K^\infty [\psi(x) - x]x^{-s-1}\,dx$$

$$\le C + s\int_K^\infty \left[x^{(1/2)+\epsilon}\right]x^{-s-1}\,dx$$

$$= C + s\int_K^\infty x^{-s-(1/2)+\epsilon}\,dx.$$

The last equation converges for $Re(s) > \frac{1}{2} + \epsilon$ and therefore so does the RHS of equation (16.9). By analytic continuation, that means the LHS of (16.9) is equal to the RHS for all $Re(s) > \frac{1}{2} + \epsilon$. And that means we cannot have $\zeta(s) = 0$ for $Re(s) > \frac{1}{2} + \epsilon$. Because our result is valid for all $\epsilon > 0$, the Riemann Hypothesis must be true. $\square$

Notes

[1] Kudryavtseva, et al [22, 28-29].

[2] Proof from Boo Rim Choe [9].

[3] Clarkson [10].

[4] Hardy and Wright [19].

[5] [11, p. 850].

[6] This proof follows an idea that (I believe) is in a handout by Steven J. Miller, Professor of Mathematics, Williams College. (See also Freitag [14, p. 193], Koch [20, p. 150] and Bonnar [7, p. 7]).

[7] We use an idea from Stein/Shakarchi [33, p. 167]. For a different proof, see Bonnar [7, p. 64].

[8] This proof based on Stein/Shakarchi [33, pp. 161-163].

[9] This proof is also based on Stein/Shakarchi [33, pp. 161-163].

[10] This proof follows ideas from Gajjar[15, Chapter 6].

[11] Used with the permission of Dr. D. R. Wilkins, School of Mathematics, Trinity College, Dublin, Ireland.

[12] This proof is based on Chandrasekharan [8, p. 88].

[13] We show here the rate which $Arg(-z) \to -\pi$ as $\delta \to 0$ for $-z = -p_1$ as it approaches the nonpositive real axis. We have $-p_1 = -\delta_x - i\delta_y = -\delta\cos(\delta) - \delta\sin(\delta)$, giving

$$\frac{-\delta\sin(\delta)}{-\delta\cos(\delta)} = \tan(\delta) \quad \text{so that} \quad \arctan\left(\frac{y}{x}\right) = \arctan\left(\frac{-\delta\sin(\delta)}{-\delta\cos(\delta)}\right) = \arctan(\tan(\delta)) = \delta.$$

Since we are in the third quadrant, that means $Arg(-z) = -\pi + \delta$. We have determined $Arg(-z)$ for $z = p_1$. For any z along the horizontal line $p_1 \to \infty$, we have $Im(z) = Im(p_1)$ and $Re(z) \geq Re(p_1)$. Thus, for any such z, $Arg(-z) = -\pi + A\delta$ for some $0 < A \leq 1$. These same calculations (except for a change of sign) apply to $Arg(-z) = \pi - A\delta$ for z along the horizontal line $\infty \to p_2$.

[14] Lemma 5.3 is based on Tom Apostol [2, pp. 77-78]. Proof of lemma 5.4, plus a nice variation of lemma 5.3 can be found at K. Chandrasekharan [8, pp. 69-70; 92-93; 97-99].

[15] Theorem 5.2 is also based on K. Chandrasekharan [8, pp. 69-70; 92-93; 97-99].

[16] See Titchmarsh [34, p. 16].

[17] The main result in this chapter is based upon Koch [20, pp. 172-176].

[18] Lemma 7.2 follows Stein/Shakarchi [33, p. 42].

[19] This proof follows Edwards [13, pp. 41-42].

[20] This first proof follows Ahlfors [1, pp. 217-218].

[21] see, e.g., Ahlfors [1, pp. 201-206].

[22] See also Titchmarsh at [34, pp. 29-30].

[23] The original proof was by von Mangoldt in 1905 [27] – later improved by Bäcklund in 1914 [4]. Our primary references for this Chapter were Titchmarsh [34, pp. 210-214] and Garrett [16].

[24] Proof based on Titchmarsh [34, p. 211].

[25] Proof based on Titchmarsh [34, p. 213] and Paul Garrett [16].

[26] See K. Chandrasekharan [8, p. 114] and Edwards [13, p. 57].

[27] Note that Edwards [13, p. 26] and Riemann use $Li(x)$ not $li(x)$. But their notation creates an ambiguity because $Li(x)$ is normally used to represent the *offset logarithmic integral*, defined as
$$Li(x) = li(x) - li(2) = \int_2^x \frac{dt}{\log t}.$$

[28] This chapter closely follows Edwards [13, 22-37] as supplemented by Li [24].

[29] Nicely explained by Edwards [13, p. 34].

[30] Much of this chapter follows Stein/Shakarchi [33, pp. 134-153].

[31] Lemma 11.7 is from Ahlfors [1, pp. 191-192].

[32] Our proof of Jensen's Formula follows Edwards [13, pp. 40-41].

[33] Lemma 11.14 is from Stein/Shakarchi [33, pp. 100-101].

[34] This section is based on Edwards [13, p. 68].

[35] This proof follows Ash/Novinger [3, pp. 154-155].

[36] We follow Ash/Novinger [3, 157-158] and Korevaar [21, 112].

[37] We will follow the approach of Newman [29], as supplemented by Korevaar [21] and Zagier [37]. We also found the following useful: Baker and Clark [6], Ash and Novinger [3] and O'Rourke [30].

[38] De la Vallée Poussin [36] was the first to show such a zero-free region. Our proof follows descriptions of his proof in Edwards [13, pp. 79-81] and Titchmarsh [34, pp. 54-55].

[39] At this point, the Edwards and Titchmarsh proofs diverge. We follow the Titchmarsh approach.

[40] There is a typo in Titchmarsh's proof. He (incorrectly) sets $(\sigma - 1) = \frac{1}{2}C/\log t$.

[41] We follow Edwards [13, pp. 81-83].

[42] We follow Edwards [13, pp. 88-91].

Bibliography

[1] Lars V. Ahlfors: *Complex Analysis*, Third Edition, McGraw-Hill, Inc., 1979.

[2] Tom M. Apostol: *Introduction to Analytic Number Theory*, First Edition, Springer-Verlag New York, Inc., 1976.

[3] R. B. Ash and W.P. Novinger: *Complex Variables*, Second Edition, 2004.

[4] R. Bäcklund: *Sur les zéros de la fonction $\zeta(s)$ de Riemann*, C.R. Acad. Sci. Paris **158**, 1979-1982 (1914).

[5] R. Bäcklund: *Uber die Nullstellen der Riemannschen Zetafunktion*, Acta Math. **41**, 345-375 (1918).

[6] Matt Baker and Dennis Clark: *The Prime Number Theorem*, Georgia Tech School of Mathematics, December 24, 2001.

[7] James Bonnar: *The Gamma Function*, (self-published), February, 2017.

[8] K. Chandrasekharan: *Lectures on The Riemann Zeta Function*, Tata Institute of Fundamental Research, Bombay, 1953.

[9] Boo Rim Choe: *An Elementary Proof of $\sum_{n=1}^{\infty} 1/n^2 = \pi^2/6$*, The American Mathematical Monthly, Vol. 94, No. 7. (Aug. - Sep., 1987), pp. 662-663.

[10] James A. Clarkson: *On the series of prime reciprocals*, Proc. Amer. Math. Soc.17(1966) 541.

[11] Philip J. Davis: *Leonhard Euler's Integral: A Historical Profile of the Gamma Function* (The American Mathematical Monthly, vol. 66 (1959), pp. 849-869).

[12] John Derbyshire: *Prime Obsession*, Joseph Henry Press, 2003.

[13] H.M. Edwards: *Riemann Zeta Function*, Dover 2001 republication, Academic Press, Inc., 1974.

[14] Eberhard Freitag and Rolf Busam: *Complex Analysis, 2nd Edition*, Springer-Verlag, 2009.

[15] Jitesh Gajjar (University of Manchester): *Advanced Mathematical Methods*, (self-published), August 23, 2012.

[16] Paul Garrett: *Counting Zeros of $\zeta(s)$*, University of Minnesota (online paper), 08-29-2013.

[17] Hadamard, J.: *Étude sur les Propriétés des Fonctions Entières et en Particulier d'une Fonction Considérée par Riemann*; J. Math. Pures Appl. [4] **9** (1893), pp. 171-215.

[18] Hadamard, J.: *Sur la distribution des zéros de la fonction $\zeta(s)$ et ses conséquences arithmétiques*; Bulletin de la S.M.F., tome 24 (1896), pp. 199-220.

[19] G. H. Hardy and E. M. Wright: *An Introduction to the Theory of Numbers*, Fifth Edition, Oxford Science Publications, 1996.

[20] Richard Koch: *Complex Variable Outline*, (online), March 16, 2016.

[21] J. Korevaar: *On Newman's Quick Way to the Prime Number Theorem*, Math. Intelligencer Volume 4, Issue 3 (1982), 108-115.

[22] E. Kudryavtseva, F. Saidak, P. Zvengrowski: *Riemann and his zeta function*, Morfismos, Vol. 9, No. 2, pp. 1–48 (2005)

[23] E. Landau: *Nouvelle démonstration pour la formula de Riemann*, Ann. Sci. École Norm. Sp. (3) **25**, 399-442 (1908).

[24] Sean Li: *Riemann's Explicit Formula*, Cornell University, May 11, 2012.

[25] Terrence P. Murphy: *Complex Analysis: a Self-Study Guide*, First Edition, Paramount Ridge Press, 2022.

[26] H. von Mangoldt: *Zu Riemann's Abhandlung 'Ueber die Anzahl der Primzahlen unter einer gegebenen Grösse'*, J. Reine Angew. Math., **114**, 255-305 (1895).

[27] H. von Mangoldt: *Zur Verteilung der Nullstellen der Riemannschen Funktion $\xi(t)$*, Math. Ann. **60**, 1-19 (1905).

[28] F. Mertens: *Über eine Eigenschaft der Riemann'schen ζ-Function*, S.-B. Akad.Wiss. Wien Math.-Natur. Kl. Abt. 2A **107**, 1429-1434 (1898).

[29] D.J. Newman: *Simple analytic proof of the prime number theorem*, Amer. Math. Monthly **87** (1980), 693-696.

[30] Ciarán O'Rourke: *The prime number theorem: Analytic and elementary proofs*, Maynooth University (thesis: online paper), 02-2013.

[31] B. Riemann: *Ueber die Anzahl der Primzahlen unter einer gegebenen Grösse* [*On the Number of Primes Less Than a Given Magnitude*] (Monatsberichte der Berliner Akademie, November 1859, 671-680) Translated by David R. Wilkins.

[32] Walter Rudin: *Principles of Mathematical Analysis, Third Edition*, McGraw-Hill, Inc., 1976.

[33] Elias M. Stein and Rami Shakarchi: *Princeton Lectures in Analysis II - Complex Analysis*, Princeton University Press, 2003.

[34] E.C. Titchmarsh: *The Theory of the Riemann Zeta Function*, Clarendon Press, Oxford, Second Edition, 1986.

[35] Charles-Jean de la Vallée Poussin: *Recherches analytiques sur la théorie des nombres (première partie)* Ann. Soc. Sci. Bruxelles **20B**, 183-256 (1896).

[36] Charles-Jean de la Vallée Poussin: *Sur la fonction $\zeta(s)$ de Riemann et le nombre des nombres premiers inférieurs a une limite donnée* Mém. Courronnés et Autres Mém. Publ. Acad. Roy. Sci., des Lettres Beaux-Arts Belg **59** (1899-1900).

[37] D. Zagier: *Newman's short proof of the Prime Number Theorem*, Amer. Math. Monthly **104** (1997), No. 8, 705-708.

Made in the USA
Columbia, SC
12 April 2022

58852313R00100